CHUA'S CIRCUIT IMPLEMENTATIONS
Yesterday, Today and Tomorrow

WORLD SCIENTIFIC SERIES ON NONLINEAR SCIENCE

Editor: Leon O. Chua
University of California, Berkeley

*To view the complete list of the published volumes in the series, please visit:
http://www.worldscibooks.com/series/wssnsa_series.shtml

WORLD SCIENTIFIC SERIES ON NONLINEAR SCIENCE
Series A Vol. 65

Series Editor: Leon O. Chua

CHUA'S CIRCUIT IMPLEMENTATIONS
Yesterday, Today and Tomorrow

Luigi Fortuna
Mattia Frasca
University of Catania, Italy

Maria Gabriella Xibilia
University of Messina, Italy

World Scientific

NEW JERSEY · LONDON · SINGAPORE · BEIJING · SHANGHAI · HONG KONG · TAIPEI · CHENNAI

Published by

World Scientific Publishing Co. Pte. Ltd.

5 Toh Tuck Link, Singapore 596224

USA office: 27 Warren Street, Suite 401-402, Hackensack, NJ 07601

UK office: 57 Shelton Street, Covent Garden, London WC2H 9HE

British Library Cataloguing-in-Publication Data
A catalogue record for this book is available from the British Library.

Cover image was taken from:
http://space.about.com/od/pictures/ig/Earth-Pictures-Gallery/The-Americas-and-Hurricane-And.--0O.htm
(permission was requested)

World Scientific Series on Nonlinear Science, Series A — Vol. 65
CHUA'S CIRCUIT IMPLEMENTATIONS
Yesterday, Today and Tomorrow

ISBN-13 978-981-283-924-4
ISBN-10 981-283-924-0

To Leon

Preface

Chua's circuit is considered a cornerstone in nonlinear electronic circuit theory. Up to now and over the last 25 years, wonderful properties of Chua's circuit have been discovered, day by day. Generations of scientists have worked on this very important device, that is the recipient of intriguing phenomena and astonishing behaviors. Even after 25 years since its discovery, Chua's circuit is still timely today.

Chua's circuit is a unique platform both for the understanding of nonlinear phenomena and for the study of experimental chaos. The more and more successful implementations of Chua's circuit allow us to have a wide scenario of electronic laboratories interested in Chua's device, both in industry and in universities.

Chua's circuit has evolved in both time and space. It is difficult to establish a geographical map of all the laboratories in the world where Chua's circuit is implemented and studied every day, due to the large number of people interested in this circuit. In our experience, the role of Chua's circuit is essential also for educational aims even in basic university courses. Chua's circuit has been present daily in our lab since 1994 and is considered a reference system to be taken into consideration to evaluate new techniques, methodologies and theories for nonlinear dynamics. The interdisciplinarities, stimulated by Chua's circuit, have led to the cooperation of groups of engineers, physicists, matter physicists, biologists. In recent years, even artists have been fascinated by it. Companies such as STMicroelectronics are investigating into the potentiality of Chua's circuit in industrial applications and a number of patents based on it have been deposited.

The book we propose reflects the wide experiences acquired together with other researchers in focalizing real circuit implementations of Chua's device.

The book is organized as follows. Chapter 1 gives an introduction to Chua's circuit. Chapter 2 deals with ways of implementing nonlinearity in the Chua's circuit. Chapter 3 presents the implementation of the Chua's circuit based on Cellular Nonlinear Networks. Starting from this implementation, two modifications allowing the generation of new nonlinear phenomena have been introduced: the first is discussed in Chapter 4, the second in Chapter 5. A programmable Chua's circuit is shown in Chapter 6. Chapter 7 presents an integrated implementation of Chua's circuit. Chapter 8 describes how to obtain a Chua's circuit with only four circuit elements. The possibility of exploiting the recent advances of organic technology to build a Chua's circuit is discussed in Chapter 9. Chapter 10 illustrates some of the applications of Chua's circuit. Chapter 11 draws the conclusions of the book.

It is worth noting that most of the circuits presented in the book can be implemented in a classical electronic lab or also at home and can be widely tested with available low-cost instrumentations or with laptop peripherals.

Usually, in our courses in Catania, we provide each student with a kit of a Chua's circuit and we make known the experimental possibility of such a circuit also in undergraduate schools.

We are indebted to the various generations of students who have collaborated with us in the experiments and in the study of the new properties and applications of Chua's circuit. We thank Leon O. Chua for his continuous encouragement in our research work. Moreover, the scientific community should be grateful to him for having offered his circuit. We have been happy to receive it from Chua himself and wish to make it known to future generations of students.

L. Fortuna, M. Frasca and M.G. Xibilia

Contents

Chapter 1

The birth of the Chua's circuit

1.1 Introduction

The Chua's circuit is the simplest electronic circuit exhibiting chaos. The circuit, shown in Fig. 1.1, consists of five elements: two capacitors, an inductor, a resistor and a nonlinear element N_R, known as the Chua's diode.

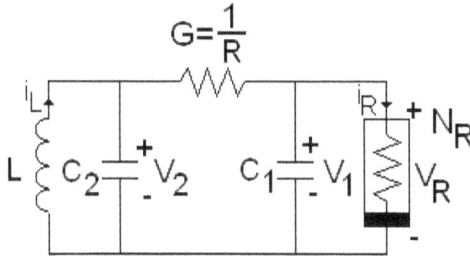

Fig. 1.1   The Chua's circuit.

While the two capacitors, the inductor and the resistor are standard electrical components, the two-terminal nonlinear resistor is an element that needs to be *ad hoc* synthesized. There are several ways to do it that will be reviewed in Chapter 2; now we will focus on its $v_R - i_R$ characteristic (i_R indicates the current flowing into it and v_R the voltage across it, as shown in Fig. 1.1). Although many nonlinear functions have been assumed for this element, in its original form it has the 3-segment piecewise-linear characteristic shown in Fig. 1.2.

This nonlinearity is fundamental to achieve an oscillatory chaotic behavior. In fact, in order to exhibit chaos, an autonomous electronic circuit must satisfy some essential criteria which are necessary (not sufficient) conditions for the appearance of chaos: the circuit must contain at least three energy-storage elements, it must contain at least one nonlinear element and it must contain at least one locally active resistor. The Chua's diode, being a nonlinear locally active resistor, allows the Chua's circuit to satisfy the last two conditions. It should be noticed at this point that the Chua's circuit does not contain too many elements beyond those strictly required for the appearance of chaos, namely only one further resistor. On the basis of such considerations, it is spontaneous to postulating the existence of a Chua's circuit based only on four elements, this is the idea which led Barboza and Chua to design the four-elements Chua's circuit that will be discussed in Chapter 8.

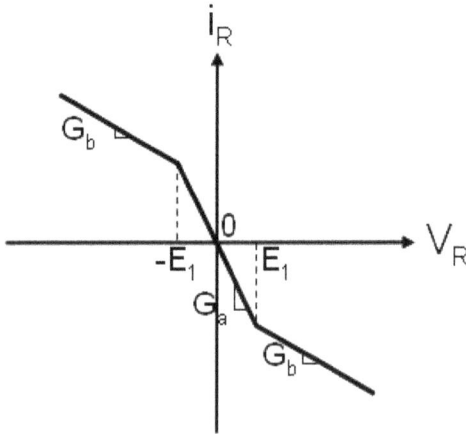

Fig. 1.2   The $v_R - i_R$ characteristic of the Chua's diode.

Let us indicate with $i_R = g(v_R)$ the current *vs.* voltage nonlinear function of the Chua's diode, with G_a the slope of the inner segment, with G_b the slope of the two outer segments and with $\pm E_1$ the breakpoints. With these assumptions the nonlinearity of the Chua's diode can be expressed as follows:

$$g(v_R) = \begin{cases} G_b v_R + (G_b - G_a)E_1, & \text{if } v_R \leq -E_1 \\ G_a v_R, & \text{if } |v_R| < E_1 \\ G_b v_R + (G_a - G_b)E_1, & \text{if } v_R \geq E_1 \end{cases} \qquad (1.1)$$

Moreover, let us indicate as v_1, v_2 and i_L the voltage across capacitor C1, the voltage across capacitor C2 and the current in the inductor L, respectively. By applying the Kirchhoff's circuit laws, the state equations of the Chua's circuit can be easily derived:

$$\begin{aligned} \frac{dv_1}{dt} &= \frac{1}{C_1}[G(v_2 - v_1) - g(v_1)] \\ \frac{dv_2}{dt} &= \frac{1}{C_2}[G(v_1 - v_2) + i_L] \\ \frac{di_L}{dt} &= -\frac{1}{L}v_2 \end{aligned} \qquad (1.2)$$

It is worth noticing that the equations governing the circuit are symmetrical with respect to the origin, *i.e.*, they are invariant under the transformation $(v_1, v_2, i_L) \rightarrow (-v_1, -v_2, -i_L)$.

For a given set of parameters (see Section 1.5.1), the circuit exhibits a chaotic strange attractor called the *double scroll strange attractor*. Figures 1.3, 1.4 and 1.5 show several projections of the attractor on the oscilloscope. Although the details of the circuit implementation used will be discussed later, here it is sufficient to say that, using such an implementation, all the three state variables are easily accessible. Figures 1.3, 1.4 and 1.5 show the projection of the double scroll strange attractor on three different phase planes, where it can be assumed that x, y and z (as rigorously defined later, in Section 1.4) are voltage signal proportional to v_1, v_2 and i_L, respectively.

From the direct inspection of the circuit (*i.e.*, by opening the capacitor and shortening the inductor), it can be observed that the DC equilibrium points are given by the intersection between the $v_R - i_R$ characteristic of the nonlinear element and the load line $-1/R$, as shown in Fig. 1.6. For the parameters fixed so that the double scroll strange attractor appears [Matsumoto *et al.* (1985)], the circuit has three equilibrium points, exactly like in Fig. 1.6. One of these equilibria is the origin, the other two are usually referred as P^+ and P^-. These two latter points are located at the center of the two holes in Fig. 1.3. A typical trajectory of the attractor rotates around one of these equilibrium points, getting further from it after each rotation until either it goes back to a point closer to the equilibrium and either repeats the process or directs toward the other equilibrium point and repeats a similar process, but around the other equilibrium point. In

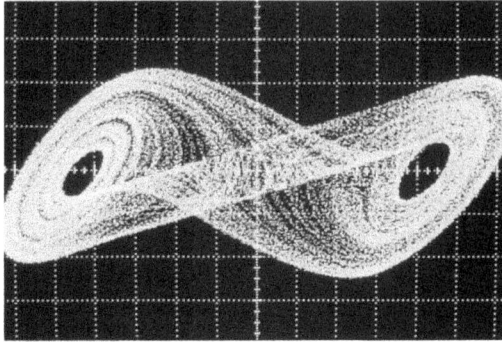

Fig. 1.3   Projection on the plane $x - y$ of the double scroll Chua's attractor. Horizontal axis: $500mV/div$; vertical axis $200mV/div$.

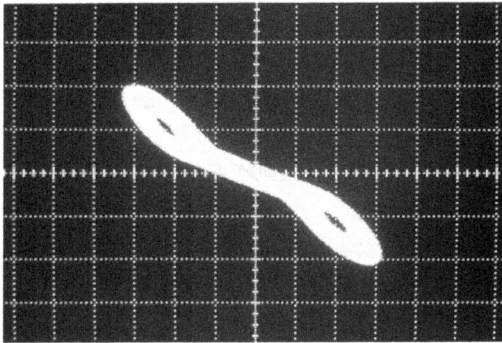

Fig. 1.4   Projection on the plane $x - z$ of the double scroll Chua's attractor. Horizontal axis: $1V/div$; vertical axis $2V/div$.

both cases the number of rotations is random. This unpredictability is one of the peculiarities of deterministic chaos.

Being a deterministic chaotic system, the Chua's circuit also exhibits the other peculiar properties of chaos [Strogatz (1994)]: it has a *high sensitivity to initial conditions*, *i.e.*, two trajectories starting from nearby close initial positions rapidly diverge and become uncorrelated, still laying on the chaotic attractor which therefore has *dense trajectories*; thanks to the *stretching and folding* mechanism the trajectories on the attractor remain confined in a bounded region of the phase space, although neighboring trajectories initially diverge in an exponential way; the attractor contains an

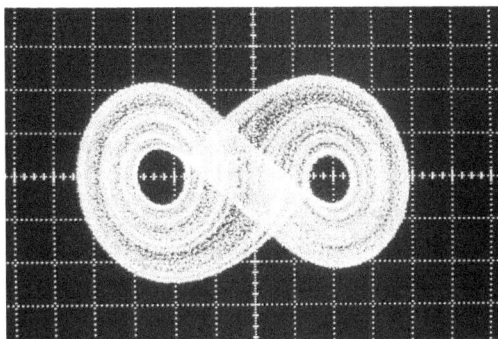

Fig. 1.5   Projection on the plane $y - z$ of the double scroll Chua's attractor. Horizontal axis: $1V/div$; vertical axis $200mV/div$.

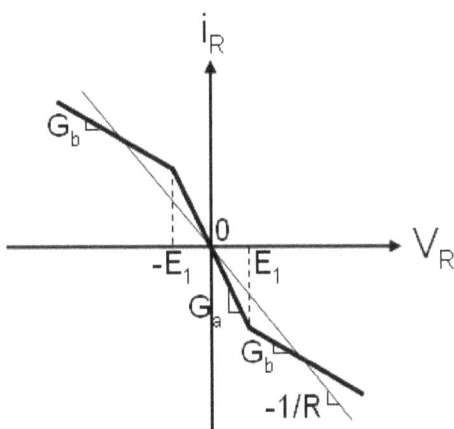

Fig. 1.6   The DC equilibrium points of the Chua's circuit can be found at the intersection between the characteristic of the nonlinear element and the load line $-1/R$.

infinite number of *unstable periodic orbits* that constitute its skeleton and that are visited during the circuit evolution; the circuit exhibits *aperiodic* oscillations and therefore has a *long-term unpredictable* behavior; the circuit has a *broadband spectrum* which derives from the fact that the circuit state variables are deterministically generated unpredictable signals.

As it has been shown, the Chua's circuit is a relatively simple circuit (in the sense that it is based on few electronic components and that its constitutive third-order equations can be easily derived) which has been rapidly become a paradigm for chaos. The main reasons for his success can be summarized in the following points:

- The Chua's circuit provides one of the simplest robust experimental proof of chaos.
- The Chua's circuit can be easily implemented in many different ways.
- A rigorous mathematical analysis can be applied to demonstrate that the Chua's circuit is chaotic.
- Many nonlinear phenomena arise in the Chua's circuit, including bifurcations, stochastic resonance, $1/f$ noise, period-adding bifurcations and so on.
- The Chua's circuit is one element of a more large family of continuous odd-symmetric piecewise-linear vector field in $\mathbb{R}^3$ to whose elements many methods and results developed for the Chua's circuit can be applied. Furthermore, a member of this family can be identified as a canonical Chua's circuit, in the sense that it is the simplest element capable of qualitatively reproducing the dynamics shown by every other member of the family. This circuit is referred to as the *canonical Chua's circuit* or the *Chua's oscillator*.
- Many applications of the Chua's circuit have been developed.

Due to these motivations, the literature on the Chua's circuit is vast and reviewing all the remarkable works on this topic is a great deal. Instead, this book aims to focus on some particular aspects of this research field, and, in particular, to those connected to the different implementations and applications of the Chua's circuit. We will limit ourselves to a brief introduction in this Chapter to some general issues on the circuit which clarify in a certain sense the above points and, at the same time, give the essential ingredients to understand the behavior of the Chua's circuit.

1.2   Genesis of the Chua's circuit

The Chua's circuit was invented in October 1983 when Leon O. Chua was visiting at the laboratory of Takashi Matsumoto in Waseda University [Chua (1992)]. At that time, there was a deep urge for reproducible chaotic circuits providing experimental evidence of chaos allowing to refute the suspect that this phenomenon was only a mathematical abstraction. This led Chua to investigate about the possibility of designing an autonomous chaotic circuit. His starting point was different from the previous attempts: instead of starting from known chaotic systems such as Lorenz or Rössler equations, he aimed at designing an electronic circuit behaving in a chaotic way.

His reasoning started from an observation. Chua noticed that in the Rössler and Lorenz system the mechanism giving rise to chaos was the same, *i.e.* the presence of at least two unstable equilibrium points (two in the Rössler system and three in the Lorenz system). Thus he decided to design a physically realizable autonomous circuit with three unstable equilibrium points (which has several advantages with respect to the case of two equilibriums points, among which the fact that it is more general and can have a nonlinearity with odd symmetry). He added the further constraint that the circuit should contain as few as possible passive elements and only one two-terminal nonlinear resistor with a piecewise-linear characteristic. Chua, then, followed a systematic step-by-step approach to design his circuit.

First of all, he determined the number of circuit elements. Keeping in mind that an autonomous dynamical system to be chaotic requires to be at least of order three, Chua fixed the elements that his circuit should contain: three energy-storage elements, one two-terminal nonlinear resistor and a number (as small as possible) of linear passive resistors.

Chua, then, selected the circuit topology. He excluded the configurations in which the three energy storage elements are either all inductors or capacitors, since they could not oscillate, and preferred the configurations with two capacitors and one inductor to the dual one with two inductors and one capacitor since high quality and tunable capacitors are less expensive than their inductive counterpart. This resulted in the configuration shown in Fig. 1.7, where N_0 is a 3-port containing only linear passive resistors.

At this point, Chua did the simplifying hypothesis that N_0 is made of only one resistor R and examined all the 8 possibilities arising from the configuration of Fig. 1.7 and reported in Fig. 1.8. The last two topologies can be immediately eliminated, since in one case (Fig. 1.8(g)) the resistor R is in parallel with the nonlinear element N_R and can be therefore included in its characteristic, and in the other case (Fig. 1.8(h)) the circuit contains two capacitors in parallel which can be obviously substituted by only one equivalent capacitor, thus resulting in a second-order dynamics.

By considering the DC equilibriums of the circuit, other four topologies shown in Fig. 1.8 can be excluded. In fact, Chua's idea was to have a non-

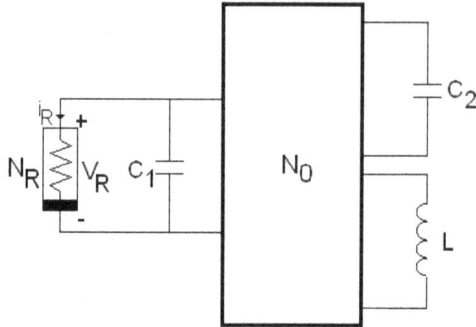

Fig. 1.7   Circuit configuration chosen by Chua [Chua (1992)] as a candidate for chaotic circuit implementation.

linearity with three segments, each one having negative slope, thus lying in the second and fourth quadrant. If one calculates the DC equilibrium circuit for the topologies of Fig. 1.8(a)-1.8(b), it can be observed that the terminals of the nonlinear element N_R (which is in parallel with the inductor) are short-circuited by the inductor. On the opposite, when the DC equilibrium circuit for the topologies of Fig. 1.8(c)-1.8(d) is derived, one finds that the terminals of the nonlinear element N_R are open. Thus, in none of these four cases the DC equilibrium points are at the intersection of a load line with finite non-zero slope: in practice, the nonlinear element N_R gives no contribution to the calculation of the DC equilibrium points, and, consequently, these topologies cannot satisfy the requirements formulated by Chua.

Although none of the two remaining topologies (*i.e.*, those shown in Fig. 1.8(e) and 1.8(f)) could be *a priori* excluded, Chua preferred that of Fig. 1.8(f) because of the presence of the LC resonant subcircuit which can provide the basis for the birth of oscillations.

The last part of the beautiful argumentation of Chua was focused on the choice of the nonlinear characteristic of N_R. Since the idea was to design a circuit with three unstable equilibriums points, the nonlinearity should have had three segments with a negative slope (remember that all the other circuit elements are passive, so N_R needs to be active in order to guarantee the instability of the equilibrium points). This observation, along with the constraint that the characteristic should be a voltage-controlled function (since it is easier to be synthesized), led to the choice of the es-

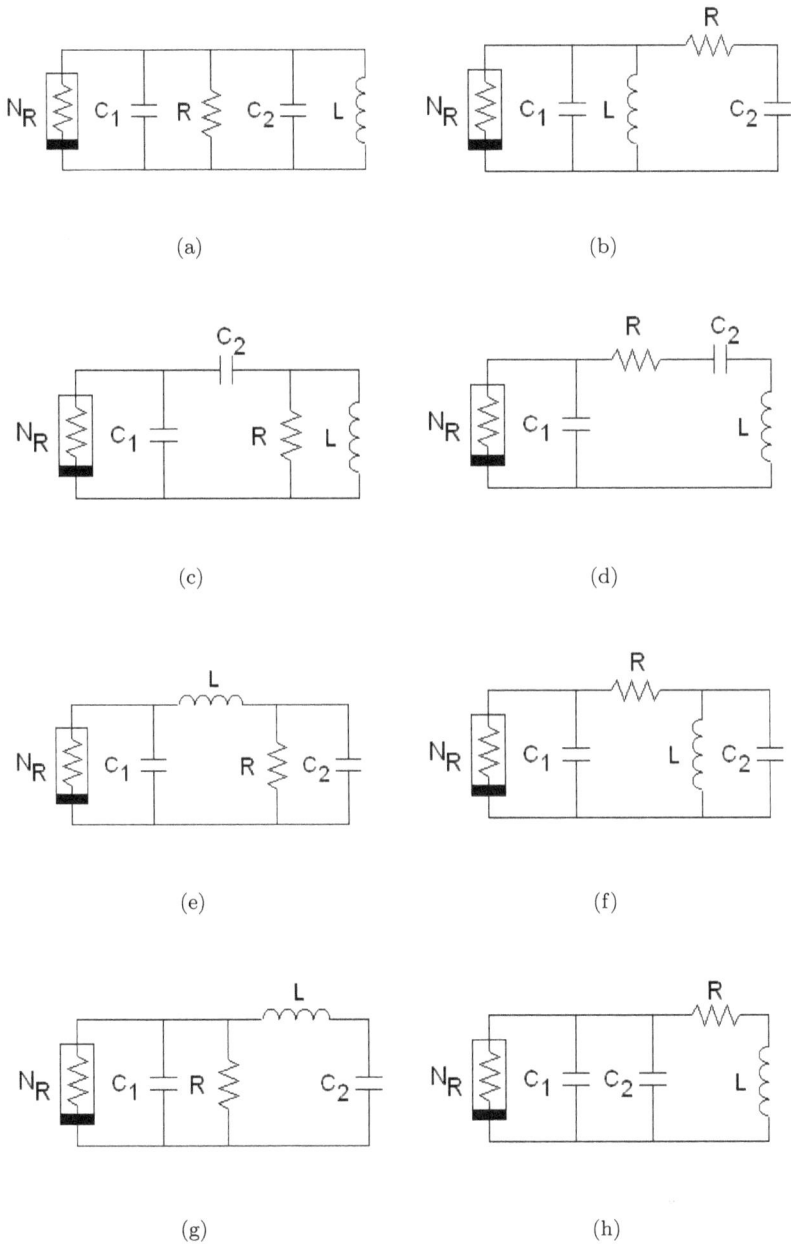

Fig. 1.8 The 8 possible circuit topologies derived from the configuration of Fig. 1.7 by assuming that N_0 is made of only one resistor.

sential nonlinearity to obtain the three unstable equilibrium points shown in Fig. 1.9(a). Taking into account the fact that this is not a realizable nonlinearity, since it is not eventually passive, Chua finally chose the nonlinear characteristic of Fig. 1.9(b), where the two outer segments do not influence the nature of the equilibrium points of the circuits, but guarantee the eventually passivity of the characteristic.

After the design of the circuit topology, its parameters were chosen by Matsumoto, Chua and Komuro [Matsumoto *et al.* (1985)] through computer simulations and taking into account that the load line should intersect at three point the three inner segments of the nonlinearity of Fig. 1.9(b). They, finally, discovered the appearence of the double scroll strange attractor, thus confirming that the circuit is effectively capable of generating chaotic behavior.

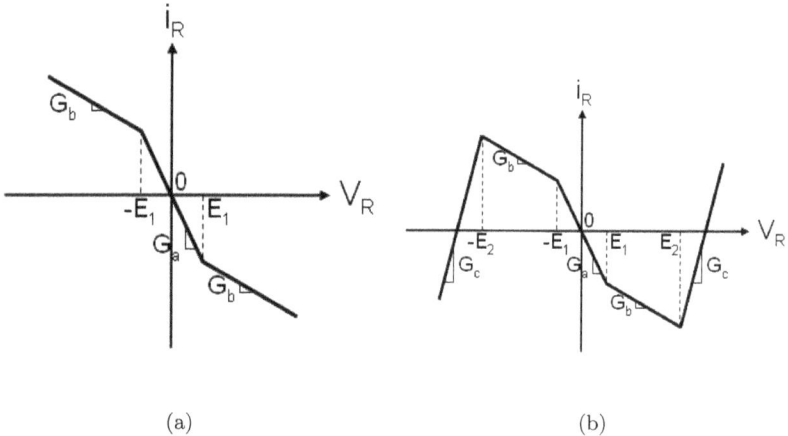

(a)   (b)

Fig. 1.9   (a) Three-segment $v_R - i_R$ characteristic of the Chua's circuit. (b) Five-segment $v_R - i_R$ characteristic of the Chua's circuit.

1.3   From RLC to Chua's circuit

As mentioned above, when Chua had to choice between the two remaining topologies with similar properties, he preferred the one containing an LC parallel. This is, in fact, the simplest mechanism providing oscillations. In [Kennedy (1993a)] a very interesting and instructive path, called by the author *evolution*, starting from the LC parallel and coming back to the Chua's circuit through circuits of increasing complexity is discussed. In this Section, the essential steps of this path are briefly reviewed with the aim of providing the reader with some further insights on the final result of this evolution.

The linear parallel RLC resonant circuit is a classical textbook example of a circuit which can oscillate. The circuit is reported in Fig. 1.10, where the state variables have been labelled in analogy with the Chua's circuit. The circuit can be described by the following state equations:

$$\begin{array}{l} \frac{dv_2}{dt} = \frac{1}{C_2}i_L - \frac{G}{C_2}v_2 \\ \frac{di_L}{dt} = -\frac{1}{L}v_2 \end{array} \tag{1.3}$$

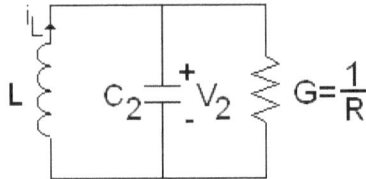

Fig. 1.10   Linear parallel RLC circuit.

The circuit has one equilibrium point at the origin, whose stability can be studied by examining the associated eigenvalues, *i.e.*, the solutions of the characteristic equation: $\lambda^2 + \frac{G}{C_2}\lambda + \frac{1}{LC_2} = 0$.

Let us consider a given initial conditions $v_2(t) = v_{20}$ and $i_L(t) = i_{L0}$. Depending on G, three cases arise. If $G > 0$, *i.e.*, if the circuit is dissipative, the origin is a stable equilibrium point and the circuit will approach it either in an overdamped (for real eigenvalues) or in an underdamped (for complex conjugate eigenvalues) way. If $G < 0$, *i.e.*, the resistor is active and supplies energy to the LC parallel, the equilibrium point is unstable and the state variables exponentially grow. Notice that this is an unrealistic case, since unbounded solutions are not feasible and, indeed, each physical resistor is eventually passive (*i.e.*, it is dissipative for large voltage across its terminal). The last case occurs for $G = 0$. In this case the equilibrium point has a pair of purely imaginary eigenvalues (*i.e.*, the circuit is undamped and the origin is neutrally stable). The energy initially stored in the inductor and in the capacitor ($v_{20} \neq 0$, $i_{L0} \neq 0$) remains constant oscillating back and forth between the two elements. The voltage across the capacitor and the current in the inductor are sinusoidal signals of the harmonic oscillator.

Although the frequency of such signals is well defined, namely $\omega = 1/\sqrt{LC_2}$, the amplitude is not a function of the circuit parameters only. In fact, it depends on initial conditions. This is, obviously, a drawback both under the viewpoint of the designer who wants a stable oscillator and under the perspective of modelling many real oscillations whose amplitude does not depend on initial conditions. This is due to the fact that the RLC circuit is not structurally stable. Therefore, something else should be added to the circuit in order to obtain stable oscillations. For this reason, since none of the three cases ($G > 0$, $G < 0$, or $G = 0$) provides any solution, a nonlinearity should be added to the circuit.

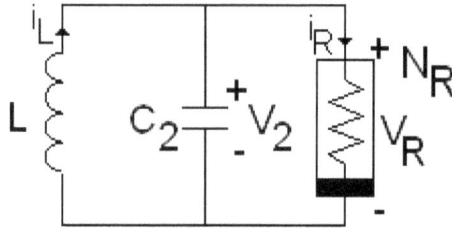

Fig. 1.11   Adding a nonlinear element in the LC circuit can generate stable oscillations.

The way in which a stable limit cycle can be generated is to add a nonlinearity to the LC parallel circuit. To this aim, introducing a 3-segment piecewise-linear resistor provides a suitable solution. Let us consider the circuit shown in Fig. 1.11 which can be obtained from that of Fig. 1.10 by substituting the linear resistor with a nonlinear one and let us assume that the nonlinear resistor has the characteristic shown in Fig. 1.12. The behavior of this circuit can be studied by taking into account that it can be decomposed in three regions in each of which it is linear. The three regions can be defined by taking into account the breakpoints of the nonlinear characteristic, *i.e.*, when $V_2 < -E_1$ (usually called D_{-1} region), when $|V_2| < E_1$ (usually called D_0 region) and when $V_2 > E_1$ (usually called D_1 region). The understanding of the behavior of the circuit in each of these regions then allows to put the pieces together and obtain a qualitative description of the whole circuit.

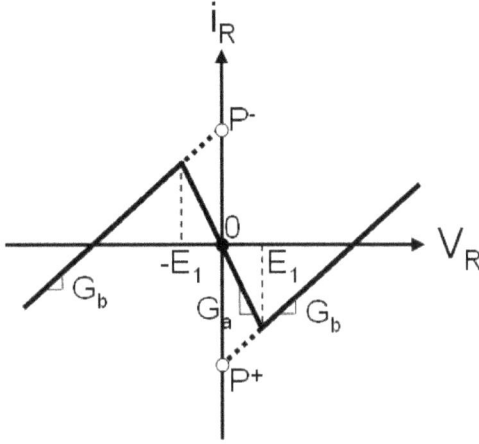

Fig. 1.12 $v - i$ characteristics of the nonlinear element used in the circuit of Fig. 1.11.

Let us notice that, in order to obtain a limit cycle G_a has to be negative (as in Fig. 1.12). In fact, when $G_a > 0$, the circuit is dissipative in each of the regions and a unique stable equilibrium point (the origin) is obtained. We refer the reader to [Kennedy (1993a)], where a more detailed description (including the case of $G_a > 0$) is given, while we focus here on $G_a < 0$.

In the region D_0, the circuit has an equilibrium point at the origin with eigenvalues given by: $\lambda_{1,2} = -\frac{G_a}{2C_2} \pm \sqrt{(\frac{G_a}{2C_2})^2 - \frac{1}{LC_2}}$. Since $C_2 > 0$, the sign of the real part of the eigenvalues is uniquely determined by G_a. For $G_a < 0$, the origin is an unstable equilibrium point. Therefore, trajectories in region D_0 are pushed away towards either D_1 or D_{-1}. The behavior in the outer regions is governed by a linear system characterized by the following eigenvalues $\lambda_{1,2} = -\frac{G_b}{2C_2} \pm \sqrt{(\frac{G_b}{2C_2})^2 - \frac{1}{LC_2}}$, which have negative real parts since $G_b > 0$. The equilibrium points of the circuit can be calculated by taking into account that at the DC equilibrum the inductor is equivalent to a short circuit and thus $v_r = 0$. This implies that the equilibrium points are at the intersection of the $v_r - i_r$ characteristic with the i_r axis. They are indicated in Fig. 1.12 with open (unstable) or closed (stable) circles. In the case under examination (*i.e.*, the outer regions) it can be observed that the equilibrium points (labelled as P^+ or P^-) lie in the D_0 region, *i.e.*, outside their regions. This implies that they are virtual equilibrium points which attract the trajectory of the system outside the region D_1 (or D_{-1}). The resulting behavior is that, when the trajectory is in D_0 it is pushed away from it, while when the trajectory is either in D_1 or D_{-1} is pushed towards the virtual equilibriums and so towards D_0. This leads to a closed trajectory, *i.e.*, a stable limit cycle.

We can notice that this case corresponds to the circuit of Fig. 1.10 with $G < 0$ and with the further consideration that the negative resistor is eventually passive, *i.e.*, that for large voltage across its terminals it has a dissipative behavior, thus resulting in a 3-segment piecewise-linear characteristic.

At this point, the next step towards the design of a chaotic circuit is to recall the initial requirements stated by Chua. He aimed to design a circuit with three unstable equilibrium points. This forces to include a resistor between the nonlinear element N_R and the LC parallel, so that the DC equilibrium is no more obtained by shortcircuiting the nonlinear resistor. Then, G_b should be fixed as negative, in order to let the two outer equilibrium point to be unstable. Finally, a third state variable (*i.e.*, a further energy-storage element) is needed. In fact, chaos requires that the autonomous circuit is at least third-order. We have already seen as these further considerations lead to the topology of the Chua's circuit.

1.4   Dimensionless Chua's equations

Equations (1.2) are usually rewritten in a more convenient form for analytical treatment. We will briefly derive these equations starting from the state equations of the circuit.

By applying the rescaling

$$
\begin{aligned}
x &= v_1/E_1 \\
y &= v_2/E_1 \\
z &= i_L/(E_1 G) \\
\tau &= tG/C_2 \\
a &= G_a/G \\
b &= G_b/G \\
\alpha &= C_2/C_1 \\
\beta &= C_2/(LG^2)
\end{aligned}
\tag{1.4}
$$

equations (1.2) can be rewritten as follows:

$$\dot{x} = \alpha[y - x - f(x)]$$
$$\dot{y} = x - y + z \qquad (1.5)$$
$$\dot{z} = -\beta y$$

where

$$f(x) = \begin{cases} bx + b - a, & \text{if } x \le -1 \\ ax, & \text{if } |x| < 1 \\ bx + a - b, & \text{if } x \ge 1 \end{cases} \qquad (1.6)$$

Usually, by defining $h(x) \triangleq x + f(x) = m_1 x + 0.5(m_0 - m_1)(|x + 1| - |x - 1|)$ (with $m_0 \triangleq a + 1$ and $m_1 \triangleq b + 1$), an equivalent form of equations (1.5) is defined:

$$\dot{x} = \alpha[y - h(x)]$$
$$\dot{y} = x - y + z \qquad (1.7)$$
$$\dot{z} = -\beta y$$

In the following, we will refer to these equations as *Chua's equations*.

1.5 Geometry of the double scroll

In this Section we will discuss some geometrical properties of the double scroll strange attractor. We will therefore refer to the set of parameters giving rise to this attractor in the Chua's circuit. In this Section we will first derive the corresponding values of the parameters appearing in the Chua's equations and, then, we will discuss the behavior of the system by examining, at first, the dynamics in each of the PWL regions and, then, by discussing the dynamics in the whole state space. More details can be found in [Matsumoto *et al.* (1985); Chua *et al.* (1986); Kennedy (1993b)].

1.5.1 *Parameter values for the Double Scroll Chua's Attractor*

The double scroll strange attractor, shown in Figs. 1.3, 1.4 and 1.5 is obtained for the following values of parameters [Matsumoto *et al.* (1985)]:

$$C_1 = 5.5nF$$
$$C_2 = 49.5nF$$
$$L = 7.07mH$$
$$R = 1.428k\Omega \quad (\text{or} \quad G = 0.7mS) \qquad (1.8)$$
$$G_a = -0.8mS$$
$$G_b = -0.5mS$$
$$E = 1V$$

Given the normalization in equations (1.4), this leads to the following values of parameters for the Chua's equations (1.7):

$$\alpha = 9$$
$$\beta = 14.2886$$
$$m_0 = -1/7 \qquad (1.9)$$
$$m_1 = 2/7$$

In the rest of this Section we will refer to this set of parameters.

1.5.2   Equilibrium points of the Chua's circuit

Since the nonlinearity of the Chua's circuit is a piecewise-linear function, the circuit can be divided into a set of separate affine regions. Analyzing the behavior of the system in each of these regions is helpful to understand the global behavior of the circuit.

In particular, if $m_0, m_1 \neq 1$, the circuit may be decomposed into three distinct affine regions:

$$D_1 \triangleq \{(x, y, z) | x \geq 1\}$$
$$D_0 \triangleq \{(x, y, z) | -1 \leq x \leq 1\} \qquad (1.10)$$
$$D_{-1} \triangleq \{(x, y, z) | x \leq -1\}$$

Notice that the planes which divide one region from the other are $x = 1$ (dividing D_1 from D_0) and $x = -1$ (dividing D_0 from D_{-1}). We refer to these planes as:

$$U_1 \triangleq D_1 \cap D_0 = \{(x, y, z) | x = 1\}$$
$$U_{-1} \triangleq D_0 \cap D_{-1} = \{(x, y, z) | x = -1\} \qquad (1.11)$$

Let us first calculate the equilibrium points of the Chua's circuit given by:

$$h(x) = 0$$
$$y = 0 \tag{1.12}$$
$$z = -x$$

In each of the three regions D_1, D_0, and D_{-1}, the system has a unique equilibrium point given by the following equations:

$$P^+ = (k, 0, k) \in D_1$$
$$\mathbf{0} = (0, 0, 0) \in D_0 \tag{1.13}$$
$$P^- = (-k, 0, -k) \in D_{-1}$$

where $k = (m_1 - m_0)/m_1$.

1.5.3 Stability of the equilibrium points

In each of the three regions D_1, D_0 and D_{-1}, the Chua's equations (1.7) are linear. Following [Matsumoto *et al.* (1985)], they can be expressed as:

$$\dot{\mathbf{x}} = \begin{cases} A(\alpha, \beta, b)(\mathbf{x} - \mathbf{k}), & \text{if } \mathbf{x} \in D_1 \\ A(\alpha, \beta, a)\mathbf{x}, & \text{if } \mathbf{x} \in D_0 \\ A(\alpha, \beta, b)(\mathbf{x} + \mathbf{k}), & \text{if } \mathbf{x} \in D_{-1} \end{cases} \tag{1.14}$$

where $\mathbf{x} = \begin{bmatrix} x & y & z \end{bmatrix}^T$, $\mathbf{k} = \begin{bmatrix} k & 0 & -k \end{bmatrix}^T$ and

$$A(\alpha, \beta, c) = \begin{bmatrix} -\alpha(c+1) & \alpha & 0 \\ 1 & -1 & 1 \\ 0 & -\beta & 0 \end{bmatrix} \tag{1.15}$$

with $c = a$ in D_0 and $c = b$ in D_1 and D_{-1}.

When Chua invented his circuit, he deliberately designed it in such a way that the three equilibrium points were unstable. So, it is not surprising that, by calculating the eigenvalues of $A(\alpha, \beta, b)$ and those of $A(\alpha, \beta, a)$, one finds in both cases at least one eigenvalue with negative real part which makes unstable the corresponding equilibrium point. However, the characteristics of the equilibrium points are different. In fact, P^+ and P^- have one negative real eigenvalue and two complex conjugate eigenvalues with positive real part, while the origin $\mathbf{0}$ has one positive real eigenvalue and two complex conjugate eigenvalues with negative real part.

More in details, if the parameters are fixed as in Eqs. (1.9), the eigenvalues associated with the equilibrium points P^+ and P^- are:

$$\gamma_p = -3.94$$
$$\sigma_p \pm j\omega_p = 0.19 \pm j3.05 \tag{1.16}$$

while those associated with the origin are:

$$\gamma_0 = 2.22$$
$$\sigma_0 \pm j\omega_0 = -0.97 \pm j2.71 \tag{1.17}$$

As will be discussed below, the eigenspaces corresponding to the eigenvalues play an important role. We, therefore, briefly discuss the equations that characterize them. Let us indicate with $E^s(\mathbf{P}^\pm)$ $(E^u(\mathbf{P}^\pm))$ the eigenspace corresponding to the real eigenvalue γ_p (to the complex conjugate eigenvalues $\sigma_p \pm j\omega_p$) and with $E^u(\mathbf{0})$ $(E^s(\mathbf{0}))$ the eigenspace corresponding to the real eigenvalue γ_p (to the complex conjugate eigenvalues $\sigma_p \pm j\omega_p$). $E^u(\mathbf{P}^\pm)$ and $E^s(\mathbf{0})$ have dimension two, while $E^s(\mathbf{P}^\pm)$ and $E^u(\mathbf{0})$ have dimension one. The eigenspaces at the origin are given by the following equations:

$$E^u(\mathbf{0}): \frac{x}{\gamma_0^2+\gamma_0+\beta} = \frac{y}{\gamma_0} = \frac{z}{-\beta} \tag{1.18}$$

$$E^s(\mathbf{0}): (\gamma_0^2 + \gamma_0 + \beta)x + \alpha\gamma_0 y + \alpha z = 0 \tag{1.19}$$

Those associated with the other two equilibrium points $\mathbf{P}^\pm$ are:

$$E^u(\mathbf{P}^\pm): \frac{x\mp k}{\gamma_p^2+\gamma_p+\beta} = \frac{y}{\gamma_p} = \frac{z\pm k}{-\beta} \tag{1.20}$$

$$E^s(\mathbf{P}^\pm): (\gamma_p^2 + \gamma_p + \beta)(x \mp k) + \alpha\gamma_p y + \alpha(z \pm k) = 0 \tag{1.21}$$

1.5.4 *General considerations on the behavior on each of the three regions D_1, D_0 and D_{-1}*

In each of the three regions, the solution of Chua's equations can be determined by solving the linear system given by Eqs. (1.14). Let us indicate with $\bar{\mathbf{x}}$ the equilibrium point in each region (*i.e.*, $\bar{\mathbf{x}} = P^+$ if $\bar{\mathbf{x}} \in D_1$, $\bar{\mathbf{x}} = \mathbf{0}$ if $\bar{\mathbf{x}} \in D_0$ and $\bar{\mathbf{x}} = P^-$ if $\bar{\mathbf{x}} \in D_{-1}$).

Moreover, let us indicate with ξ_r the eigenvector associated with the real eigenvalue γ (either γ_p or γ_0 depending on the region under examination) and with η_r and η_i the real and the imaginary part of the eigenvectors associated with $\sigma \pm \omega$ (either $\sigma_p \pm \omega_p$ or $\sigma_0 \pm \omega_0$).

The solution of Chua's equations in a given region can be written in a general form as:

$$\mathbf{x}(t) = \bar{\mathbf{x}} + \mathbf{x}_r(t) + \mathbf{x}_c(t) \tag{1.22}$$

with $\mathbf{x}_r(t) = c_1 e^{\gamma t}\xi_r$ and $\mathbf{x}_c(t) = c_c e^{\sigma t}[\cos(\omega t + \phi_c)\eta_r - \sin(\omega t + \phi_c)\eta_r]$. c_1, c_c and ϕ_c are real constants which depend on the initial conditions.

Depending on the given region, such terms may tend towards zero or grow exponentially. For instance, if we consider region D_1, then $\gamma = \gamma_p < 0$ and $\sigma = \sigma_p > 0$. Thus, $\mathbf{x}_r(t)$ tends to zero, while $\mathbf{x}_c(t)$ spirals away from P^+. Instead, if we consider region D_0 the opposite holds: $\gamma = \gamma_0 > 0$, $\sigma = \sigma_0 < 0$, $\mathbf{x}_r(t)$ exponentially grows, and $\mathbf{x}_c(t)$ spirals towards the origin.

Another important consideration is that the eigenspaces E^u and E^s are invariant, *i.e.*, if a trajectory starts from one of them, then it will remain on it for all the time. This implies that the eigenspaces associated with the complex conjugate eigenvectors (which are the planes $E^u(\mathbf{P}^\pm)$ and $E^s(\mathbf{0})$) cannot be crossed by the trajectory $\mathbf{x}(t)$. This consideration is fundamental to understand qualitatively the dynamics of the double scroll strange attractor.

1.5.5 Qualitative description of the dynamics in D_1 (or D_{-1})

Let us consider, at first, a trajectory starting from an initial condition in D_1 (for symmetry all the considerations can be applied to the a trajectory starting from D_{-1}). In this region, the equilibrium point P^+ is associated with a real negative eigenvalue and a pair of conjugate eigenvalues with positive real part. The eigenspace spanned by the two complex eigenvectors is the plane $E^u(\mathbf{P}^\pm)$. As discussed above, this plane cannot be crossed by a trajectory, which will therefore remain above the plane if the initial state is above the plane or, otherwise, will remains below the plane if the initial point is below the plane. The trajectory can be decomposed into its components $\mathbf{x}_r(t)$ and $\mathbf{x}_c(t)$. Since $\gamma_p < 0$, the component in the direction of the real eigenvalue will tend towards the equilibrium point P^+. On the other hand, since $\sigma_p > 0$ the component along the plane $E^u(\mathbf{P}^\pm)$ will spiral away from P^+. The whole resulting trajectory, starting from a point above or below the plane $E^u(\mathbf{P}^\pm)$, will, therefore, be rapidly flattened on the plane and then spiral away from the equilibrium point. At some point, this trajectory, characterized by an increasing radius, will cross the boundary between D_1 and D_0 (*i.e.*, the plane U_1) and enter the region D_0.

1.5.6   Qualitative description of the dynamics in D_0

In the region D_0, the equilibrium point of Chua's equation is the origin, and is associated with a positive real eigenvalue and a pair of conjugate eigenvalues with negative real part. In this region the plane spanned by the two eigenvalues is $E^s(\mathbf{0})$. This plane is also an invariant of the flow and thus cannot be crossed by a trajectory, so that trajectories starting above (below) this plane will remain for all time above (below) it. A trajectory in D_0 has two components: the first (associated with the complex conjugate eigenvalues with negative real part) will spirals towards the origin, while the second (associated with the positive real eigenvalue) will grow exponentially. As a consequence, the whole trajectory from a point above the plane spiralling with a decreasing radius is pushed away from the origin above the plane and in the direction of the eigenvector associated to the real eigenvalues.

1.5.7   The double scroll attractor

Having in mind the dynamics of the Chua's equations in each of the regions in which they can be decomposed, a qualitative description of the trajectory of the double scroll attractor can be given. Let us take into consideration a trajectory starting from D_1. As discussed before, this trajectory will be flattened on $E^u(\mathbf{P}^\pm)$ and then follow a helix of increasing radius until it crosses U_1 and enters the region D_0. At this point two cases may occur. If the trajectory crosses U_1 and enters the region D_0 above the plane $E^s(\mathbf{0})$, then it will be pushed away above the plane (and so towards U_1) and enter again the region D_1. Otherwise, if the trajectory crosses U_1 below the plane $E^s(\mathbf{0})$, it will be pushed towards U_{-1} and the region D_{-1}. Therefore, depending on how the trajectory will cross U_1, it will be forced back to D_1 or to D_{-1}.

This mechanism causes the typical trajectory to rotate around P^+ or P^- for a certain time and then to go back to the equilibrium or be pushed towards the other in an irregular way and accounts for the sensitive dependence on initial conditions. Two trajectories starting from two very close points in D_1 may evolve in a complete different manner depending on where they will cross the plane U_1. In fact, if one trajectory will cross U_1 slightly above $E^s(\mathbf{0})$ and the other slightly below $E^s(\mathbf{0})$, their evolutions will be totally different.

The geometry of the double scroll strange attractor derives from all the considerations dealt with in this Section. Figures 1.13 and 1.14 shows two different three dimensional views of the double scroll strange attractor along with the planes $E^u(\mathbf{P}^+)$, $E^s(\mathbf{0})$, $E^u(\mathbf{P}^-)$, U_1 and U_{-1}. The axes have been oriented so that the axis x is vertical and different colors have been used for the planes shown. For the sake of clarity, in Fig. 1.14 $E^u(\mathbf{P}^+)$ and $E^u(\mathbf{P}^-)$ have been omitted.

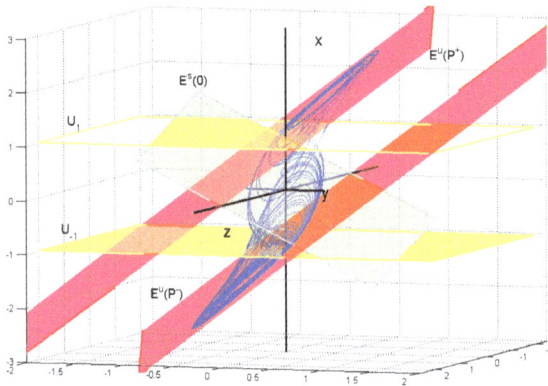

Fig. 1.13   The double scroll Chua's attractor.

As it can be noticed in the outer regions (D_1 and D_{-1}) the attractor is flattened on either $E^u(\mathbf{P}^+)$ or $E^u(\mathbf{P}^-)$. Furthermore, in Fig. 1.14 it seems clear how $E^s(\mathbf{0})$ bisects the attractor separating those trajectories coming from D_1 and going back in it, from those that, crossing U_1 from below the plane, go towards region D_{-1} (the same applies to the trajectories coming from D_{-1} and directed to D_1, if crossing the plane from above, or back to D_{-1}, if crossing the plane from below).

1.5.8   *Rigorous proof of chaos in the Chua's circuit*

The chaotic nature of the double scroll strange attractor has been rigorously proven in [Chua *et al.* (1986)]. The mathematical proof involves the

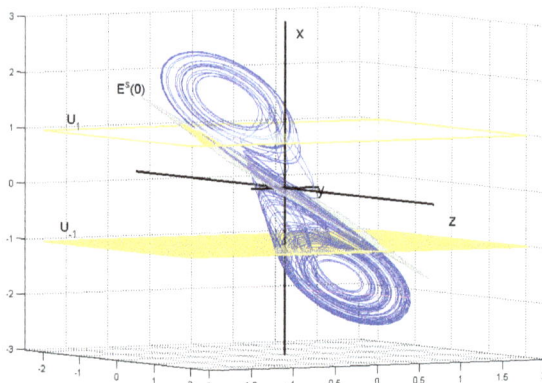

Fig. 1.14   The double scroll Chua's attractor.

analytical calculation of several Poincaré maps which go beyond the scope of this introduction. We only briefly discuss what are the guidelines of the rigorous proof presented in [Chua *et al.* (1986)]. The chaotic behavior of the double scroll strange attractor is proven by means of the Shilnikov's theorem [Guckenheimer and Holmes (1983); Shilnikov (1965); Arneodo *et al.* (1982)], that states that, if in a continuous piecewise-linear vector field associated with a third-order autonomous system the origin is an equilibrium point having a pair of complex eigenvalues $\sigma + j\omega$ with $\sigma < 0$ and a real eigenvalue $\gamma > 0$ and $|\sigma| < \gamma$, and there exists a homoclinic orbit through the origin, then a countable set of horseshoes, which are a fingerprint of chaos, appear if the vector field is infinitesimally perturbed. In [Chua *et al.* (1986)] the Shilnikov's theorem is indeed applied to a class of piecewise-linear vector fields including the original Chua's equations. More in details, the proof is based on the following main steps:

- The class of piecewise-linear vector fields including the original Chua's equations, called the double scroll family, is defined.
- An explicit normal form for the vector fields belonging to the double scroll family is derived.
- Poincaré maps and analytical expressions for the associated half-return maps are calculated.
- The half-return maps are used to prove the existence of the

Shilnikov-type homoclinic orbit and, thus, to demonstrate that the hypotheses of the Shilnikov's theorem hold.

- Furthermore, the derived Poincaré maps are used to draw a bifurcation diagram and characterize the birth and death of the double scroll strange attractor.

1.6 Bifurcations of the Chua's circuit

The behavior of the circuit with respect to its parameters has been extensively studied both in simulations and in experiments in many papers [Chua *et al.* (1986); Madan (1993); Kahlert and Chua (1987); Yang and Liao (1987)]. In particular, in [Chua *et al.* (1986)] a rigorous analysis of bifurcation phenomena is discussed and the complete bifurcation diagram with respect to the parameters α and β is reported. This detailed analysis is beyond the scope of this Chapter, where we will limit ourselves to discuss typical scenarios of the complex bifurcation diagram of the Chua's circuit.

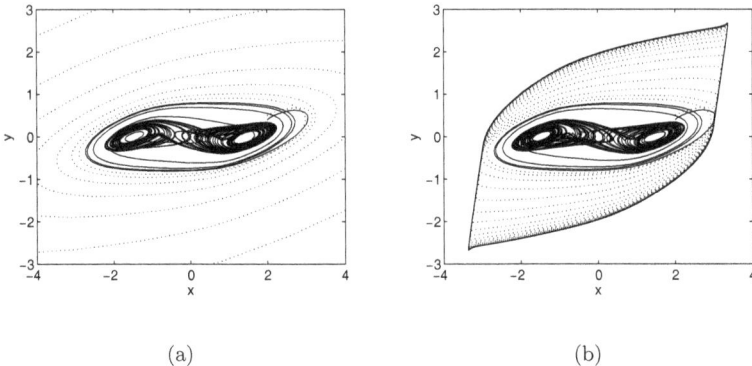

(a)   (b)

Fig. 1.15   Differences between the effects of the three and five segment nonlinearity. (a) When the three segment nonlinearity is considered, the behavior of the Chua's equations may be unstable. (b) When the five segment nonlinearity is considered, the double scroll strange attractor coexists with a stable external limit cycle. The initial conditions are $(2, 0.4, 0)$ (continuous line) and $(2, 0.5, 0)$ (dashed line).

First of all, a remark on the nonlinearity of the Chua's circuit is needed. In the real circuit, since each real nonlinear resistor with negative slope is eventually passive, *i.e.*, it has positive slope for large voltage values, the five segment nonlinear characteristic of Fig. 1.9(b) is implemented. However, for many purposes, *e.g.*, for the analysis of the circuit behavior presented in Section 1.5, it is enough to take into account the three segment nonlinearity. This is no more true when the behavior of the Chua's circuit at different initial conditions is investigated or a bifurcation analysis is carried on. In fact, when the three segment nonlinearity is considered, for large initial conditions the behavior of the system may be unstable, which is clearly not the case of the real circuit. When the five segment nonlinearity is taken into account, it can be demonstrated that the double scroll attractor coexists with a stable external limit cycle and that an unstable saddle-type periodic orbit separates the basins of attraction of the two attractors.

In Fig. 1.15 we compare two simulations starting from two different initial conditions for both the two nonlinearities. As it can be noticed, when the five segment nonlinearity is considered, the double scroll attractor coexists with a stable external limit cycle, while the same initial condition leads to instability when the external segment with positive slopes are not taken into account.

In the real circuit, the coexistence of the two attractors clearly appears, as for the same set of parameters different initial conditions may lead to the double scroll attractor shown in Fig. 1.16(a) or to the external limit cycle shown in Fig. 1.16(b).

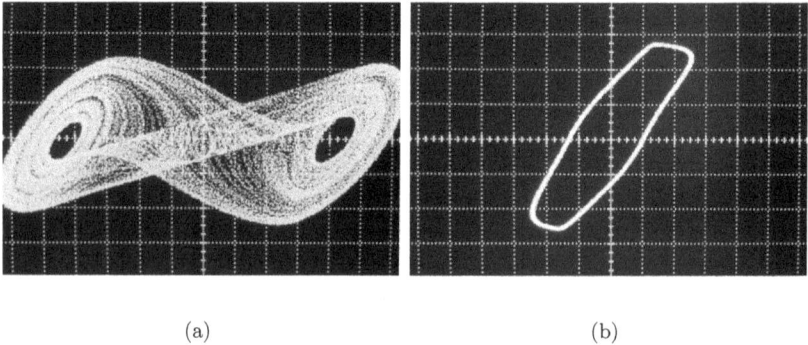

(a)   (b)

Fig. 1.16   Two different initial conditions in the Chua's circuit may lead to one of the different coexisting attractors: (a) double scroll attractor (phase plane: $x - y$, horizontal axis: $500mV/div$; vertical axis $200mV/div$); (b) external limit cycle (phase plane: $x - y$, horizontal axis: $2V/div$; vertical axis $5V/div$).

A typical bifurcation parameter of the Chua's circuit is the resistance R

or equivalently the parameter α in the Chua's equations. For large values of R the equilibrium points P^+ and P^- are stable. Starting from this condition and decreasing R (or equivalently starting from a small value of α and increasing it) the first bifurcation that can be observed is the loss of stability of the equilibrium points P^+ and P^- through a Hopf bifurcation and, thus, the birth of two symmetric stable limit cycles. One of these symmetric limit cycle can be observed in Fig. 1.17(a).

Decreasing further the parameter α a sequence of period-doubling bifurcations can be observed. Period-2, period-4 and period-8 limit cycles are shown in Fig. 1.17(b), 1.17(c) and 1.17(d), respectively. Figure 1.17(e) is a magnification of Fig. 1.17(d). This sequence of period doubling bifurcations lead to chaos through the well-known period-doubling route-to-chaos. The chaotic attractor that can be observed is shown in Fig. 1.17(f). This attractor is confined to the two regions D_1 and D_0 and is referred to either as *single scroll attractor* or *Rössler screw-type attractor* for its resemblance to the structure of the Rössler attractor. Since the circuit is symmetric, a mirrored single scroll attractor lies in the regions D_{-1} and D_0. For further increasing of α these two distinct attractors grow in size until they collide, giving birth to the double scroll attractor which spans all the three regions D_1, D_0 and D_{-1}, as shown in Fig. 1.16(a).

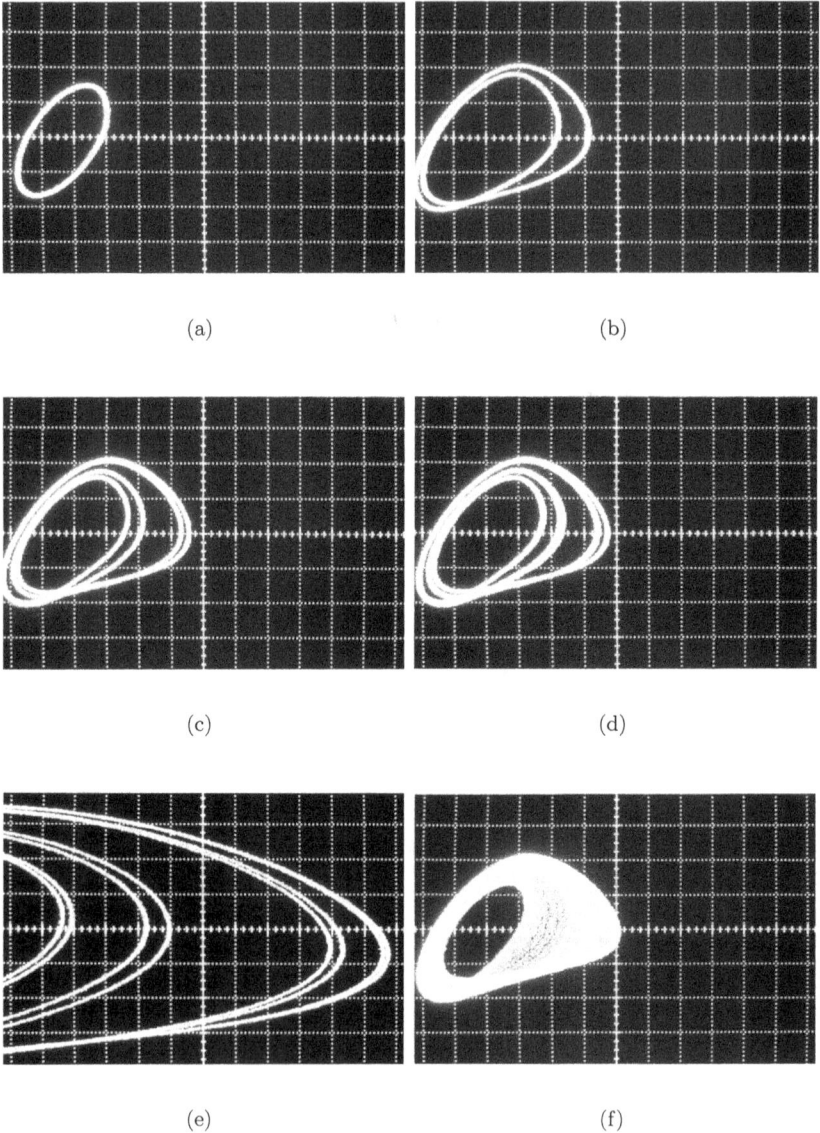

(a)   (b)

(c)   (d)

(e)   (f)

Fig. 1.17 Period-doubling bifurcations leading to chaos: (a) period-1 limit cycle; (b) period-2 limit cycle; (c) period-4 limit cycle; (d)period-8 limit cycle; (e) zoom of period-8 limit cycle; (f) single scroll chaotic attractor. Phase plane: $x - y$, Horizontal axis: $500mV/div$; vertical axis $200mV/div$ except for (e): Horizontal axis: $100mV/div$; vertical axis $100mV/div$.

Another typical scenario in the region of parameters α and β is the death of the double attractor, which occurs, for instance, when α is further increased. The double scroll attractor grows in size, until it collides with the unstable saddle-type periodic orbit and disappears. The result is that only the stable external limit cycle is observed for such parameters.

Between the regions of chaotic behaviors, stable periodic orbits of periodic window type can also be observed. One of such orbit (a period-3 limit cycle) is shown in Fig. 1.18.

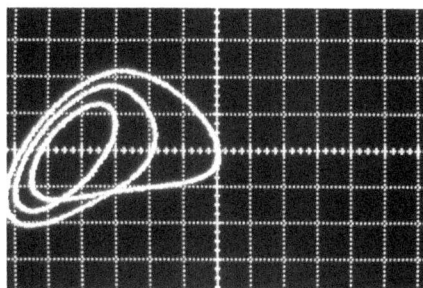

Fig. 1.18   Period-3 limit cycle. Phase plane: $x - y$, Horizontal axis: $500mV/div$; vertical axis $200mV/div$.

The bifurcation scenario described above has been illustrated through some experimentally observed trajectories generated from the implementation based on Cellular Nonlinear Networks described in Chapter 3. It can be numerically reproduced by varying α in the range of $\alpha = 6.5$ (stable equilibrium points) to $\alpha = 11$ (stable external limit cycle) as shown in Fig. 1.19. Figure 1.20 reports the bifurcation diagram obtained with respect to β, also showing a great variety of complex behaviors.

1.7   The Chua's oscillator

Starting from the Chua's circuit a much larger family of vector fields with continuous odd-symmetric piecewise-linear nonlinearities can be defined [Chua (1993)]. The great advantage of defining/identifying such class of circuits is that many of the theories and methods developed for the Chua's circuit can be directly applied to any member of this family. Furthermore, canonical circuits, *i.e.*, circuits which are able to reproduce all the dynamics of any member of the family, can be identified. Their realization allows any

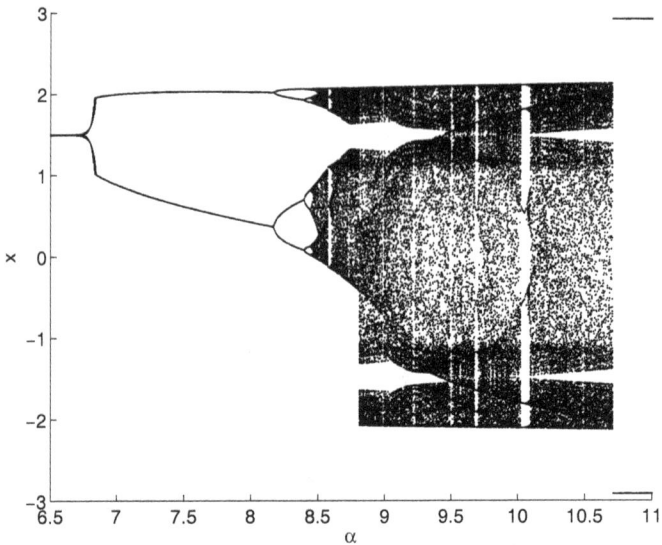

Fig. 1.19   Bifurcation diagram with respect to α. The other parameters are as in Eq. (1.9).

dynamics of the class to be observed in a unique real circuit. The general properties that a circuit should have to belong to the Chua's circuit family are listed in the following definition.

Definition of the Chua's circuit family

A circuit defined by the state equation $\dot{\mathbf{x}} = f(\mathbf{x})$ with $\mathbf{x} \in \mathbb{R}^3$ belongs to the Chua's circuit family C if and only if:

(1) $f(\cdot)$ is continuous;
(2) $f(\cdot)$ is odd-symmetric, *i.e.*, $f(-\mathbf{x}) = -f(\mathbf{x})$;
(3) the state space can be partitioned by two parallel boundary planes U_1 and U_{-1} into three regions D_1, D_0 and D_{-1}.

Under such hypothesis and the further assumption that the boundary planes are defined by $x = 1$ (U_1) and $x = -1$ (U_{-1}) as in Section 1.5, each member of the family C is described by the following equations:

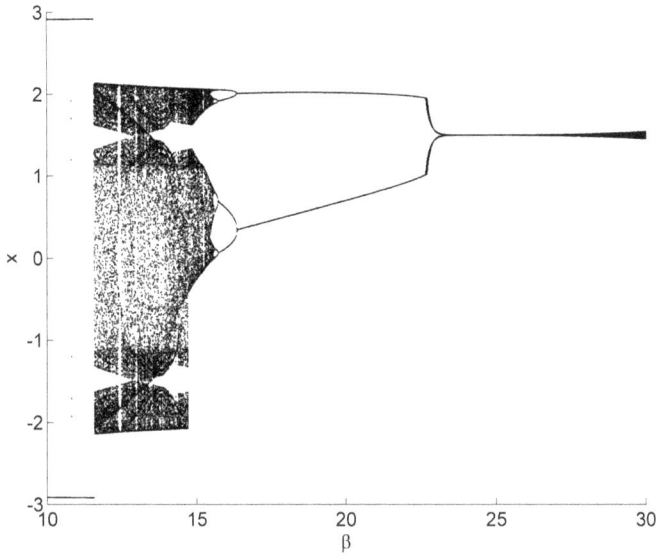

Fig. 1.20   Bifurcation diagram with respect to β. The other parameters are as in Eq. (1.9).

$$\dot{\mathbf{x}} = \begin{cases} \mathbf{A}\mathbf{x} + \mathbf{b}, & \text{if } x \geq 1 \text{ or } x \leq -1 \\ \mathbf{A_0}\mathbf{x} & \text{if } |x| \geq 1 \end{cases} \tag{1.23}$$

with $\mathbf{A} = \begin{bmatrix} a_{11} & a_{12} & a_{13} \\ a_{21} & a_{22} & a_{23} \\ a_{31} & a_{32} & a_{33} \end{bmatrix}$ and $\mathbf{b} = \begin{bmatrix} b_1 \\ b_2 \\ b_3 \end{bmatrix}$ in regions D_1 and D_{-1} and

$\mathbf{A_0} = \begin{bmatrix} \alpha_{11} & \alpha_{12} & \alpha_{13} \\ \alpha_{21} & \alpha_{22} & \alpha_{23} \\ \alpha_{31} & \alpha_{32} & \alpha_{33} \end{bmatrix}$ in region D_0.

Although 21 parameters appear in Eqs. (1.23), taking into account the constraint that the vector field is continuous, only 12 parameters can be arbitrarily fixed. For this reason the Chua's circuit family is a 12-parameter family.

Given this family, the question that one may ask (and that Chua, Lin and other researchers did [Chua and Lin (1990); Chua (1993); Xu (1987)]) is whether it is possible to synthetize a circuit representative of the whole family, *i.e.*, able to qualitatively reproduce the dynamics of any member of the family. If it exists, this circuit will be defined as a *canonical* member of the family.

In order to illustrate how finding such a canonical circuit is effectively possible, the starting point is the concept of linearly conjugacy. Two circuits (or vector fields) of the Chua's circuit family are linearly conjugate if and only if their corresponding eigenvalues in each region are identical.

Two linearly conjugate circuits are equivalent, showing a qualitative identical behavior. In practice, although the attractors exhibited by two linearly conjugate circuits may seem different, a linear transformation from the state space of one circuit to that of the second circuit can be defined in such a way that the two attractors are exactly the same. In other words the phase portrait of one circuit can be smoothly deformed into the phase portrait of the other circuit.

On the basis of such considerations, a canonical Chua's circuit can be defined as a circuit whose parameters can be chosen so that it is linearly conjugate to each member of the family. A realization of such circuit will allow to implement any dynamics of the Chua's circuit family by appropriately choosing the parameters of the canonical circuit. It will be shown in the following that different circuits satisfy this property; among these circuits, the one having the simplest structure is selected.

We will first show that the Chua's circuit is not a canonical member of the family. Let us indicate with μ_1, μ_2 and μ_3 the eigenvalues in the inner region and with ν_1, ν_2 and ν_3 those of the outer region (since the circuit is odd-symmetric, in fact, the eigenvalues in D_1 and those in D_{-1} are the same). As introduced above, we can select the parameters of a canonical circuit so that it is linearly conjugate of an arbitrarily chosen circuit of the family. Therefore, given a set of eigenvalues μ_1, μ_2, μ_3, ν_1, ν_2 and ν_3, it is possible to choose the parameters of the canonical circuit so that it has such eigenvalues. We will show that this is not possible for the Chua's circuit. Let us consider the characteristic polynomial associated with the inner region D_0:

$$(s - \mu_1)(s - \mu_2)(s - \mu_3) = s^3 - p_1 s^2 + p_2 s - p_3 \qquad (1.24)$$

and that associated with the outer regions

$$(s - \nu_1)(s - \nu_2)(s - \nu_3) = s^3 - q_1 s^2 + q_2 s - q_3 \qquad (1.25)$$

where

$$p_1 = \mu_1 + \mu_2 + \mu_3$$
$$p_2 = \mu_1\mu_2 + \mu_2\mu_3 + \mu_1\mu_3$$
$$p_3 = \mu_1\mu_2\mu_3$$
$$q_1 = \nu_1 + \nu_2 + \nu_3 \qquad (1.26)$$
$$q_2 = \nu_1\nu_2 + \nu_2\nu_3 + \nu_1\nu_3$$
$$q_3 = \nu_1\nu_2\nu_3$$

Although the eigenvalues can be complex, the coefficients p_1, p_2, p_3, q_1, q_2 and q_3 of the characteristic polynomials are real numbers. Thus, it is more convenient to state the problem of finding a circuit linearly conjugate to a given one in terms of such parameters.

Given a set of eigenvalues μ_1, μ_2, μ_3, ν_1, ν_2 and ν_3, if we want to find a Chua's circuit linearly conjugate to this, we should find a set of parameters C_1, C_2, L, G, G_a and G_b such that the Chua's circuit has the given eigenvalues. To do this, we can calculate the characteristic polynomials in the inner and in the outer regions in terms of C_1, C_2, L, G, G_a and G_b and equal them to equations (1.24) and (1.25).

In the inner region D_0, the state equations of the Chua's circuit (1.2) are linear and assume the following form:

$$
\begin{bmatrix} \frac{dv_1}{dt} \\ \frac{dv_2}{dt} \\ \frac{di_L}{dt} \end{bmatrix} =
\begin{bmatrix} -\frac{G+G_a}{C_1} & \frac{G}{C_1} & 0 \\ \frac{G}{C_2} & -\frac{G}{C_2} & \frac{1}{C_2} \\ 0 & -\frac{1}{L} & 0 \end{bmatrix} \qquad (1.27)
$$

In the outer regions, the state equations assume the following form:

$$
\begin{bmatrix} \frac{dv_1}{dt} \\ \frac{dv_2}{dt} \\ \frac{di_L}{dt} \end{bmatrix} =
\begin{bmatrix} -\frac{G+G_b}{C_1} & \frac{G}{C_1} & 0 \\ \frac{G}{C_2} & -\frac{G}{C_2} & \frac{1}{C_2} \\ 0 & -\frac{1}{L} & 0 \end{bmatrix} \pm
\begin{bmatrix} (G_b - G_a)E_1 \\ 0 \\ 0 \end{bmatrix} \qquad (1.28)
$$

where the plus sign applies in the region D_1, and the minus sign in the region D_{-1}.

At this point the characteristic polynomial in the inner region can be calculated:

$$p(s) = s^3 + \left(\frac{G+G_a}{C_1} + \frac{G}{C_2} \right) s^2 + \left(\frac{GG_a}{C_1C_2} + \frac{1}{C_2L} \right) s + \frac{G+G_a}{C_1C_2L} \qquad (1.29)$$

In order to choose the parameters of the Chua's circuit to match the desired set of eigenvalues in the inner region, *i.e.*, μ_1, μ_2 and μ_3, equation (1.29) is compared with equation (1.24), so that it is obtained:

$$\frac{G+G_a}{C_1} + \frac{G}{C_2} = -p_1$$
$$\frac{GG_a}{C_1C_2} + \frac{1}{C_2L} = p_2 \qquad (1.30)$$
$$\frac{G+G_a}{C_1C_2L} = -p_3$$

By calculating the characteristic polynomial in the outer region

$$p(s) = s^3 + \left(\frac{G+G_b}{C_1} + \frac{G}{C_2}\right)s^2 + \left(\frac{GG_b}{C_1C_2} + \frac{1}{C_2L}\right)s + \frac{G+G_b}{C_1C_2L} \qquad (1.31)$$

and comparing it with equation (1.25), an expression similar to (1.30) is obtained for the parameters q_1, q_2 and q_3:

$$\frac{G+G_b}{C_1} + \frac{G}{C_2} = -q_1$$
$$\frac{GG_b}{C_1C_2} + \frac{1}{C_2L} = q_2 \qquad (1.32)$$
$$\frac{G+G_b}{C_1C_2L} = -q_3$$

Equations (1.30) and (1.32) represent a system of six linear equations in six unknown. The condition for which it admits a nontrivial (*i.e.*, nonzero) solution is that:

$$(p_2 - q_2)(p_3 - q_3) = (p_1 - q_1)(p_3q_1 - q_3p_1) \qquad (1.33)$$

Furthermore, we have to assume that $p_1 \neq q_1$, $p_2 \neq q_2$ and $p_3 \neq q_3$.

Equation (1.33) represents a strong constraint on the eigenvalues that can be realized by the Chua's circuit. Therefore, not all the members of the Chua's circuit family can be synthesized by the Chua's circuit. This leads to the conclusion that the Chua's circuit is not a canonical member of the family. From a theoretical point of view, the analysis reported in [Chua and Lin (1990)] explains how in practice the six unknown parameters do not provide six degrees of freedom but only five. In fact, for the so-called impedance scaling property, if for a given set of eigenvalues $\bar{C}_1$, $\bar{C}_2$, $\bar{G}$, $\bar{G}_a$, $\bar{G}_b$ and $\bar{L}$ constitute a solution of equations (1.30) and (1.32), $k\bar{C}_1$, $k\bar{C}_2$, $k\bar{G}$, $k\bar{G}_a$, $k\bar{G}_b$ and $\bar{L}/k$ also are a solution. One of such parameters can therefore arbitrarily fixed, thus decreasing the number of free parameters. Therefore, the number of parameters of the canonical circuit should be at least seven. We will show that adding a further resistor to the Chua's circuit leads to a canonical member of the family.

The circuit shown in Fig. 1.21 is obtained just by adding a resistor to the Chua's circuit in series with the inductor. This circuit is called the Chua's oscillator and, as shown below, is a canonical member of the Chua's circuit family. The Chua's oscillator can be described by the following equations:

$$\begin{aligned}\frac{dv_1}{dt} &= \frac{1}{C_1}[G(v_2 - v_1) - g(v_1)]\\\frac{dv_2}{dt} &= \frac{1}{C_2}[G(v_1 - v_2) + i_L]\\\frac{di_L}{dt} &= -\frac{1}{L}(v_2 + R_0 i_L)\end{aligned} \tag{1.34}$$

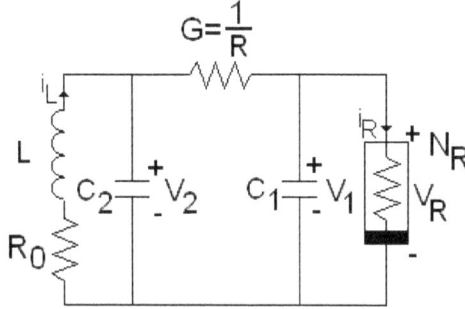

Fig. 1.21   The Chua's oscillator.

The characteristic polynomial in the inner region is

$$\begin{aligned}p(s) = s^3 &+ \left(\frac{G+G_a}{C_1} + \frac{G}{C_2} + \frac{R_0}{L}\right)s^2 +\\&+ \left(\frac{GG_a}{C_1 C_2} + \frac{G+G_a}{C_1 L}R_0 + \frac{GR_0}{C_2 L} + \frac{1}{C_2 L}\right)s + \frac{R_0 GG_a + G + G_a}{C_1 C_2 L}\end{aligned} \tag{1.35}$$

and that in the outer regions is

$$\begin{aligned}p(s) = s^3 &+ \left(\frac{G+G_b}{C_1} + \frac{G}{C_2} + \frac{R_0}{L}\right)s^2 +\\&+ \left(\frac{GG_b}{C_1 C_2} + \frac{G+G_b}{C_1 L}R_0 + \frac{GR_0}{C_2 L} + \frac{1}{C_2 L}\right)s + \frac{R_0 GG_b + G + G_b}{C_1 C_2 L}\end{aligned} \tag{1.36}$$

By comparing equations (1.24) and (1.25) with equations (1.35) and (1.36) it can be obtained:

$$\begin{aligned}\frac{G+G_a}{C_1} + \frac{G}{C_2} + \frac{R_0}{L} &= -p_1\\\frac{GG_a}{C_1 C_2} + \frac{G+G_a}{C_1 L}R_0 + \frac{GR_0}{C_2 L} + \frac{1}{C_2 L} &= p_2\\\frac{R_0 GG_a + G + G_a}{C_1 C_2 L} &= -p_3\\\frac{G+G_b}{C_1} + \frac{G}{C_2} + \frac{R_0}{L} &= -q_1\\\frac{GG_b}{C_1 C_2} + \frac{G+G_b}{C_1 L}R_0 + \frac{GR_0}{C_2 L} + \frac{1}{C_2 L} &= q_2\\\frac{R_0 GG_b + G + G_b}{C_1 C_2 L} &= -q_3\end{aligned} \tag{1.37}$$

Equations (1.37) represent a system of six linear equations with seven unknowns, that can be solved by fixing a suitable value to one of the circuit parameters (C_1):

$$
\begin{aligned}
C_1 &= 1 \\
C_2 &= \frac{k_2}{k_3^2} \\
L &= -\frac{k_3^2}{k_4} \\
R &= -\frac{k_3}{k_2} \\
R_0 &= -\frac{k_1 k_3^2}{k_2 k_4} \\
G_a &= -p_1 - \left(\frac{p_2 - q_2}{p_1 - q_1}\right) + \frac{k_2}{k_3} \\
G_b &= -q_1 - \left(\frac{p_2 - q_2}{p_1 - q_1}\right) + \frac{k_2}{k_3}
\end{aligned}
\tag{1.38}
$$

with

$$
\begin{aligned}
k_1 &= -p_3 + \left(\frac{q_3 - p_3}{q_1 - p_1}\right)\left(p_1 + \frac{p_2 - q_2}{q_1 - p_1}\right) \\
k_2 &= p_2 - \left(\frac{q_3 - p_3}{q_1 - p_1}\right) + \left(\frac{p_2 - q_2}{q_1 - p_1}\right)\left(\frac{p_2 - q_2}{q_1 - p_1} + p_1\right) \\
k_3 &= \left(\frac{p_2 - q_2}{q_1 - p_1}\right) - \frac{k_1}{k_2} \\
k_4 &= -k_1 k_3 + k_2\left(\frac{p_3 - q_3}{p_1 - q_1}\right)
\end{aligned}
\tag{1.39}
$$

If one of the following conditions

$$
\begin{aligned}
&p_1 - q_1 = 0 \\
&p_2 - \left(\frac{q_3 - p_3}{q_1 - p_1}\right) + \left(\frac{p_2 - q_2}{q_1 - p_1}\right)\left(\frac{p_2 - q_2}{q_1 - p_1} + p_1\right) = 0 \\
&\left(\frac{p_2 - q_2}{q_1 - p_1}\right) - \frac{k_1}{k_2} = 0 \\
&-k_1 k_3 + k_2\left(\frac{p_3 - q_3}{p_1 - q_1}\right) = 0
\end{aligned}
\tag{1.40}
$$

holds, then solution (1.38) cannot be obtained. However, since conditions (1.40) define a set of measure zero, it can be concluded that the unfolding Chua's circuit is able to realize any member of the Chua's circuit family except for a set of measure zero. Furthermore, if we want to implement a given circuit belonging to this set of measure zero, since this circuit is continuous, we can consider a perturbation of it and implement the perturbed circuit. From the considerations discussed in this Section, the conclusion stated in [Chua (1994)] appears evident: "The Chua's oscillator is structurally the simplest and dynamically the most complex member of the Chua's circuit family".

Among the other equivalent canonical members of the Chua's circuit family, we only mention the circuit introduced in [Chua and Lin (1990)] and reported in Fig. 1.22.

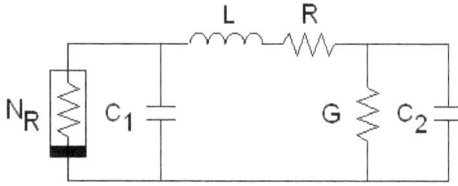

Fig. 1.22   Another canonical member of the Chua's circuit family.

Given the set of eigenvalues μ_1, μ_2, μ_3, ν_1, ν_2 and ν_3 of the member of the Chua's circuit family to be synthesized, the parameters of the canonical circuit of Fig. 1.22 can be found by using the following relationships:

$$
\begin{aligned}
C_1 &= 1 \\
G_a &= -p_1 + \tfrac{p_2 - q_2}{p_1 - q_1} \\
G_b &= -q_1 + \tfrac{p_2 - q_2}{p_1 - q_1} \\
L &= \left[p_2 + \left(\tfrac{p_2 - q_2}{p_1 - q_1} - p_1 \right) \left(\tfrac{p_2 - q_2}{p_1 - q_1} \right) - \tfrac{p_3 - q_3}{p_1 - q_1} \right]^{-1} \\
R &= -L \left(\tfrac{p_2 - q_2}{p_1 - q_1} + K \right) \\
C_2 &= L^{-1} \left[\tfrac{p_3 - q_3}{p_1 - q_1} + K \left(K + \tfrac{p_2 - q_2}{p_1 - q_1} \right) \right]^{-1} \\
G &= K C_2
\end{aligned}
\tag{1.41}
$$

with $K = -L \left[p_3 + \tfrac{G_a (p_3 - q_3)}{C_1 (p_1 - q_1)} \right]$. In this case the constraint $p_1 \neq q_1$ should be satisfied, which again defines a set of zero measure.

We notice that for some values of the eigenvalues set to be implemented, the parameters of the Chua's oscillator may assume negative values. The implementation of such negative circuit parameters (for instance, a negative capacitor) requires specific circuital solutions that will be examined in Chapter 2.

The Chua's oscillator can be described by the following dimensionless equations:

$$
\begin{aligned}
\dot{x} &= k\alpha [y - h(x)] \\
\dot{y} &= k (x - y + z) \\
\dot{z} &= k (-\beta y - \gamma z)
\end{aligned}
\tag{1.42}
$$

with $\gamma = C_2 / L / G$ and $k = \mathrm{sgn}(C_2 / G)$. Thousand of chaotic attractors have been found for different parameters of the Chua's oscillator [Bilotta and Pantano (2008)]. We show only some examples in Figs. 1.23 and 1.24.

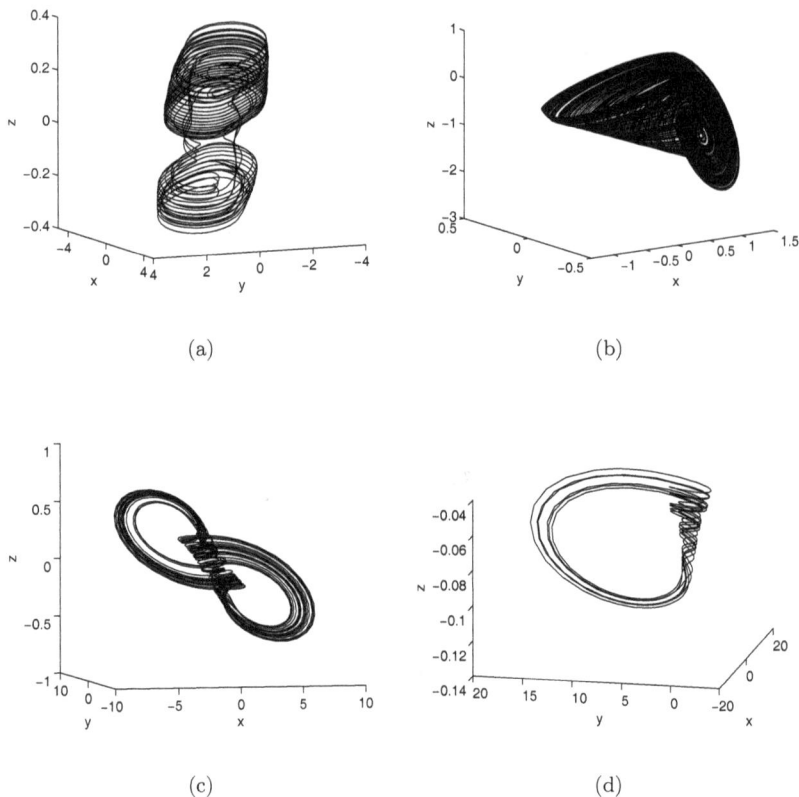

Fig. 1.23   Some of the attractors exhibited by the Chua's oscillator for different parameter values: (a) $m_0 = 1.1690$, $m_1 = 0.5230$, $\alpha = -1.3011$, $\beta = -0.0136$, $\gamma = -0.0297$, $k = 1$; (b) $m_0 = 0.02$, $m_1 = -1.4$, $\alpha = -75.0188$, $\beta = 31.25$, $\gamma = -3.1250$, $k = 1$; (c) $m_0 = 0.7562$, $m_1 = 0.9575$, $\alpha = -1.5601$, $\beta = 0.0156$, $\gamma = 0.1581$, $k = -1$; (d) $m_0 = 0.9059$, $m_1 = 0.9998$, $\alpha = -1.0870$, $\beta = 9.6899e - 005$, $\gamma = 0.0073$, $k = -1$.

To further illustrate the richness of the dynamics of the Chua's circuit, we give a list of some of the many nonlinear phenomena that have been discovered in the Chua's circuit:

- Standard *routes to chaos* (including period doubling and torus breakdown) have been observed in the Chua's circuit. We have shown in Section 1.6 a cascade of period-doubling bifurcations leading to chaos.
- *Stochastic resonance* has been observed in the Chua's circuit [An-

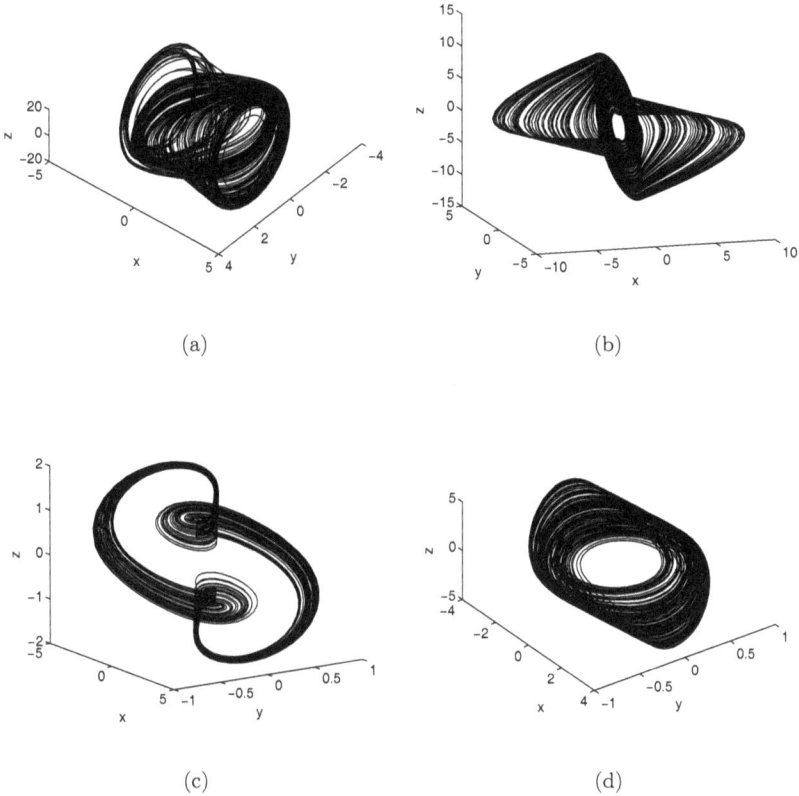

Fig. 1.24 Some of the attractors exhibited by the Chua's oscillator for different parameter values: (a) $m_0 = -1.7640$, $m_1 = 1.1805$, $\alpha = 3.7092$, $\beta = 24.0964$, $\gamma = -0.8602$,
$k = 1$; (b) $m_0 = -1.4000$, $m_1 = 0.0200$, $\alpha = -75.0188$, $\beta = 31.7460$, $\gamma = -3.1746$,
$k = 1$; (c) $m_0 = -0.1429$, $m_1 = 0.2858$, $\alpha = -4.087$, $\beta = -2$, $\gamma = 0$, $k = 1$; (d)
$m_0 = 0.2952$, $m_1 = -0.1460$, $\alpha = 8.342$, $\beta = 11.925$, $\gamma = 0$, $k = 1$.

ishchenko *et al.* (1992); Gomes *et al.* (2003)]. The Chua's circuit in
a regime close to the crisis bifurcation that leads the two symmetrical
spiral attractors to collide into the double scroll strange attractor may
exhibit stochastic resonance [Anishchenko *et al.* (1992)]. In fact, in
such region of parameters the circuit may exhibit chaos-chaos intermittency that, in the presence of a small sinusoidal signal, may result in
an increase of the signal-to-noise ratio of the output signal.

• *Signal amplification via chaos* has been also observed when the circuit operates in a spiral attractor regime [Halle *et al.* (1992)].

- Close to the bifurcation between the regime of spiral attractors and that of the double scroll strange attractor, another interesting phenomenon has been observed: the appearance of noise with $1/f$ power spectrum [Anishchenko *et al.* (1993)]. This suggests that the Chua's circuit can be used as $1/f$ noise generator.

- The reversal of period-doubling cascades, *i.e.*, *antimonotonicity*, in the Chua's circuit has been theoretically predicted [Dawson *et al.* (1992)] and experimentally observed [Kocarev *et al.* (1993)].

- Beyond period-doubling in which the oscillation period doubles, in the Chua's oscillator sequences of limit cycles with oscillation periods increasing by consecutive integers have been also observed [Pivka and Spany (1993); Mayer-Kress *et al.* (1993)]. This phenomenon is called *period adding*.

- Other nonlinear phenomena, such hyperchaos or spiral waves, have been observed in higher-order generalizations of the Chua's circuit as shown in the next Section.

1.8   Generalizations of the Chua's circuit

The main generalization of the Chua's circuit, *i.e.*, the Chua's oscillator, has already been discussed. However, the Chua's circuit has been generalized in many other directions. One of this is the inclusion of an external forcing which makes the circuit non-autonomous. Another generalization is to higher dimensions (for instance adding a further energy storage element, in order to obtain a fourth-order system exhibiting hyperchaos). Further studies have led to the use of a smooth nonlinearity instead of the piecewise function. Finally, another important generalization is the use of a piecewise function with more than five segment which leads to a generalization of the double scroll strange attractor. This Section is devoted to a brief overview of these generalizations.

1.8.1   *Sinusoidal forcing in the Chua's circuit*

Chaotic behavior in non-autonomous circuits, *i.e.*, in systems driven by some external signal, has been observed in many circuits: the Duffing equation [Duffing (1918)] and the Van der Pol oscillator [der Pol and der Mark (1927)] are only two well-known examples. Although the peculiarity of the

Chua's circuit is that it constitutes the first demonstration of chaotic behavior in an autonomous circuit, the effects of a sinusoidal forcing in the Chua's circuit have been also investigated.

According to the scheme proposed in [Murali and Lakshmanan (1992)] and shown in Fig. 1.25, a sinusoidal forcing may be introduced in series with a further inductor constituting a new branch in parallel with C_2 and the original inductor of the Chua's circuit. Under such assumptions, the state equations of the system become:

$$\begin{aligned}
\frac{dv_1}{dt} &= \frac{1}{C_1}[G(v_2 - v_1) - g(v_1)] \\
\frac{dv_2}{dt} &= \frac{1}{C_2}[G(v_1 - v_2) + i_{L1} + i_{L2}] \\
\frac{di_{L1}}{dt} &= -\frac{1}{L1}(v_2 + u(t)) \\
\frac{di_{L2}}{dt} &= -\frac{1}{L2}v_2
\end{aligned} \tag{1.43}$$

with the usual nonlinearity $g(v_1)$. The experimental investigation of this circuit carried on in [Murali and Lakshmanan (1992)] reveal a great variety of bifurcation sequences. In particular, period-adding bifurcations, quasi-periodicity, hysteresis and intermittent behaviors have been observed.

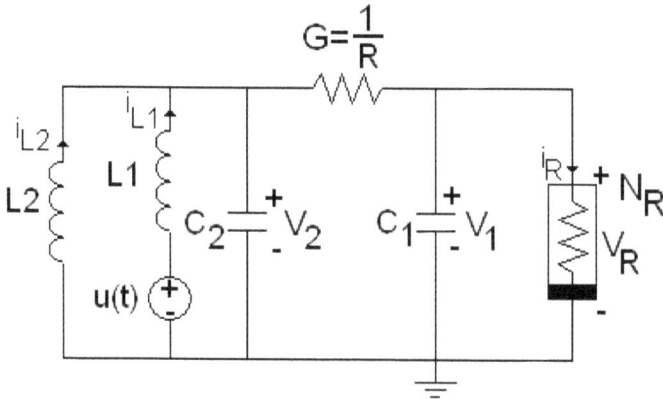

Fig. 1.25 Scheme of the driven Chua's circuit.

Another interesting result is the possibility of controlling many of these phenomena by adding a further sinusoidal generator in series with the previous one. In [Murali and Lakshmanan (1993)], for example, a second sinusoidal forcing with a different frequency is used. Experiments carried on increasing the amplitude of the second forcing demonstrate that a small amplitude is sufficient to induce drastic changes in the behavior of the system. In particular, starting from a chaotic behavior in absence of the second forcing, Murali and Lakshmanan demonstrate that periodic orbits (for instance of period-3) can be stabilized adding a sinusoidal term with a small amplitude.

This research line finally led to the ideation of a much simpler non-autonomous circuit, at the core of which is the Chua's diode, constituting thus the nonlinearity of the system [Murali *et al.* (1994b,a); Lakshmanan and Murali (1995)]. The circuit, known as the *Murali-Lakshmanan-Chua circuit*, is shown in Fig. 1.26. It consists of three linear element (L, C and R), the Chua's diode and a sinusoidal forcing and may exhibit many different complex phenomena observed in the driven Chua's circuit.

The Chua's circuit driven by a periodic forcing may also show other complex nonlinear phenomena. In particular, in [Liu (2001)] the authors demonstrate the existence of *strange nonchaotic attractors*, *i.e.*, attractors which have a fractal geometry but nonpositive Lyapunov exponents [Grebogi *et al.* (1984)].

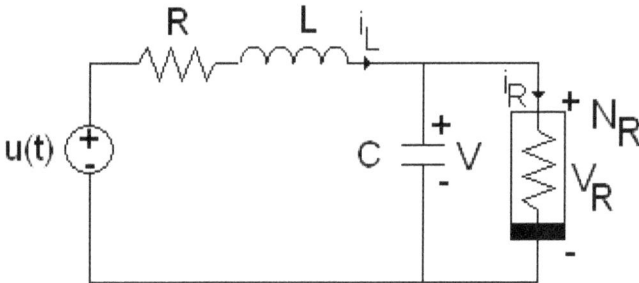

Fig. 1.26   The Murali-Lakshmanan-Chua circuit [Murali *et al.* (1994b)].

1.8.2 *Hyperchaos in the Chua's circuit*

The Chua's circuit can also generate hyperchaotic behavior. Hyperchaos is a chaotic behavior in which two or more Lyapunov exponents are positive. As such to be observed in an autonomous circuit it needs at least a fourth-order system. In [Cannas and Cincotti (2002); Cincotti and Stefano (2004)] it is shown that two bidirectionally coupled Chua's circuits can generate hyperchaos. A nonlinear resistor is used to couple the two circuits.

The experimental observation of hyperchaos in coupled Chua's circuits is reported in [Kapitaniak *et al.* (1994)], where five identical Chua's circuits are unidirectionally coupled. Hyperchaotic attractors have been experimentally observed both in open and closed chains.

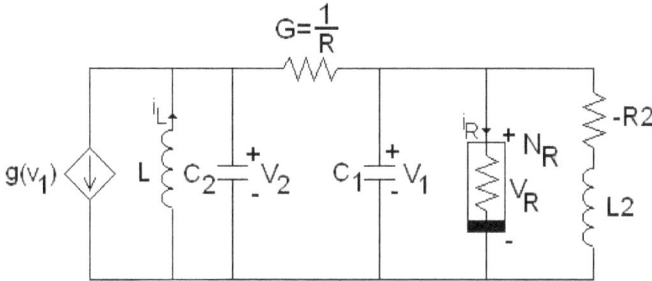

Fig. 1.27 The hyperchaotic Chua's circuit [Barboza (2008)].

In all the examples reported above, hyperchaos is generated by coupling two or more Chua's circuits. Instead, Barboza [Barboza (2008)] added two branches to the Chua's circuit and demonstrated that hyperchaos can be generated by the Chua's circuit. In fact, in order to obtain hyperchaos from the Chua's circuit, a modification is needed: at least a further energy storage element should be introduced so that the new circuit is at least of fourth order. In the scheme proposed by Barboza only two new branches are added to the original topology of the Chua's circuit. In particular, he added the series of two linear elements (a negative resistance and an inductor) in parallel with C1 and a voltage-controlled current source in parallel with C2. The voltage-controlled current source is described by a piecewise-linear function with the same breakpoints of the Chua's diode. The circuit is shown in Fig. 1.27. The dimensionless equations governing such circuit are:

$$\begin{aligned}
\dot{x} &= \alpha[y - x + w - f(x)] \\
\dot{y} &= x - y + z - g(x) \\
\dot{z} &= -\beta y \\
\dot{w} &= \gamma(w - x)
\end{aligned} \tag{1.44}$$

where $f(x) = bx + 0.5(a - b)(|x + 1| - |x - 1|)$ as usual and $g(x) = 0.5c(|x + 1| - |x - 1|)$. An example of the hyperchaotic behavior obtained for the following set of parameters ($a = -11/7$, $b = -1/7$, $c = 15$, $\alpha = 10$, $\beta = 20$, $\gamma = 1.0$) is given in Fig. 1.28.

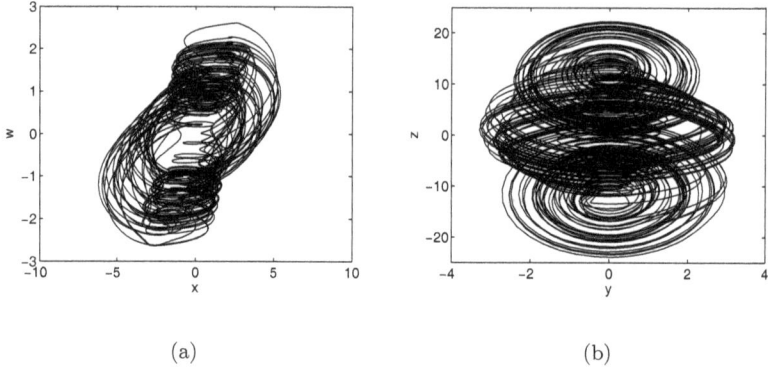

(a)   (b)

Fig. 1.28   Projection of the attractor of the hyperchaotic Chua's circuit into the plane $x - w$ (a) and $y - z$ (b).

Further generalizations to higher dimensions are dealt with in [Lukin (1993); Sharkovsky *et al.* (1993)], where the RLC subcircuit is substituted by a coaxial cable or by a lossless transmission line. Finally, many works have investigated higher order systems made of elementary units which are Chua's circuits. Among those works, we mention studies on one- or two-dimensional arrays of Chua's circuits [Perez-Munuzuri *et al.* (1992, 1993a,b,c,d); Nekorkin *et al.* (1995, 1996a,b); Nekorkin and Chua (1993); Perez-Munuzuri *et al.* (1995); Yang and Chua (2001)], showing how a large variety of nonlinear phenomena may arise in such systems: spiral waves, autowaves, solitary waves, Turing patterns and spatio-temporal chaos are only a few representative examples. Recently, studies on three-dimensional reaction-diffusion systems made of Chua's circuits have revealed possible relationships with art and neuroscience [Arena *et al.* (2005)].

1.8.3   *The Chua's circuit with a smooth nonlinearity*

One of the generalizations of the Chua's circuit investigated in several papers (see for instance [Hartley and Mossayebi (1993a,b); Altman (1993a); Khibnik *et al.* (1993a,b)]) is the use of a smooth nonlinearity instead of the piecewise function. The model can be described by the Chua's equations

$$\dot{x} = \alpha[y - \varphi(x)]$$
$$\dot{y} = x - y + z \qquad (1.45)$$
$$\dot{z} = -\beta y$$

with $\varphi(x) = c_0 x^3 + c_1 x$, where $c_0 = 1/16$ and $c_1 = -1/6$. This cubic polynomial nonlinearity well approximates the smooth nonlinearity observed in the real circuit, although the piecewise function has the great advantage to make possible the analysis described in the previous Sections. The dynamics of the smooth model and the piecewise circuit are similar, but there are also some differences such as the appearance of both supercritical and subcritical Andronov-Hopf bifurcations in the smooth model [Khibnik *et al.* (1993b)]. Furthermore, continuation technique and classical bifurcation theory can be applied to the Chua's circuit with a smooth nonlinearity.

1.8.4 *The n-scroll chaotic attractor*

In [Suykens and Vandewalle (1993)] it is shown how, by modifying the characteristic of the nonlinear resistor, the Chua's circuit can be generalized to a circuit exhibiting more complex attractors. The nonlinearity is modified by introducing additional breakpoints. The attractor generated in this way is a generalization of the double scroll strange attractor called *n-scroll chaotic attractor* or a *multiscroll chaotic attractor* with $n = 1, 2, 3, \ldots$. In this context the double scroll strange attractor corresponds to the 1- double scroll.

The multiscroll attractor is thus generated by adding more segments to the characteristic of the nonlinear resistor. In particular, the following nonlinear function [Suykens *et al.* (1997)] should be adopted in equations (1.7) to obtain a $n-$scroll attractor:

$$h(x) = m_{2q-1} + \tfrac{1}{2} \sum_{i=1}^{2q-1} (m_{i-1} - m_i)(|x + c_i| - |x - c_i|) \qquad (1.46)$$

where q is a natural number, c_i are the breakpoints of the nonlinearity and m_i the slope of i-th segment. Examples of a $3-$ and $4-$scroll chaotic attractor are given in Fig. 1.29. In Fig. 1.29(a) the following parameters have been chosen: $\alpha = 9$; $\beta = 14.286$; $m_0 = 0.9/7$; $m_1 = -3/7$; $m_2 = 3.5/7$; $m_3 = -2.4/7$; $c1 = 1$; $c_2 = 2.15$; $c_3 = 4$; while in Fig. 1.29(a) the following parameters have been used: $\alpha = 9$; $\beta = 14.286$; $m_0 = -1/7$; $m_1 = 2/7$; $m_2 = -4/7$; $m_3 = 2/7$; $c_1 = 1$; $c_2 = 2.15$; $c_3 = 3.6$.

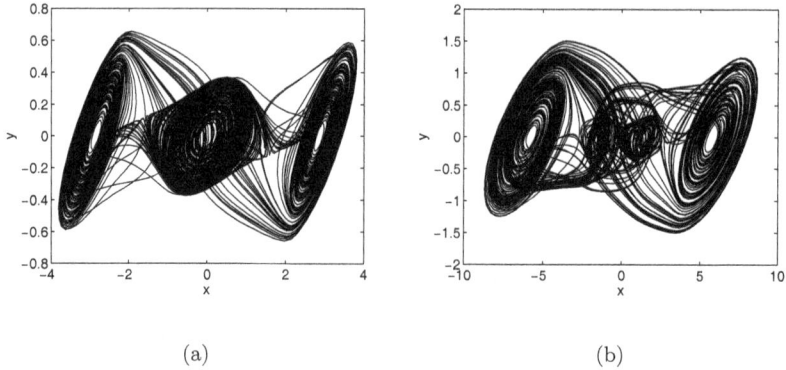

(a)   (b)

Fig. 1.29   (a) 3−scroll chaotic attractor. (b) 4−scroll chaotic attractor.

Experimental confirmation of n−scroll attractors generated by generalized Chua's circuits have been reported in [Arena *et al.* (1996a,b)] and [Yalcin *et al.* (2000)]. After their introduction, multiscroll chaotic attractors have been extensively investigated, even in relation with circuit topologies different from that of the Chua's circuit. Starting from Chua's circuit, implementations based on it or other different approaches have been reported. A part the above mentioned generation of n−scrolls from the generalized Chua's circuit, other approaches have been developed: sine-function-based circuits [Tang *et al.* (2001)], Cellular Nonlinear Networks [Arena *et al.* (1996a); Suykens and Chua (1997)], nonlinear transconductor [Ozoguz *et al.* (2002)], stair function method [Yalcin *et al.* (2002)], hysteresis series switching [Lü *et al.* (2004b)] and saturated function series [Lü *et al.* (2004a); Hulub *et al.* (2006)]. Many of these approaches (see for instance [Yalcin *et al.* (2002); Lü *et al.* (2004b,a); Hulub *et al.* (2006)]) allow to obtain also two-dimensional or three-dimensional grids of scrolls.

1.9   Control of the Chua's circuit

The term *chaos control* refers to different control problems such as the stabilization of equilibrium points or the stabilization of some periodic orbit and, in general, is the process which brings order into disorder, *i.e.*, that suppresses chaos [Chen (1993); Ogorzalek (1993b, 1995); Boccaletti *et al.* (2000)]. On the opposite, when chaotic behavior is intentionally created by using control, this is referred as *antichaos control*.

Many different techniques have been applied either to chaos control or antichaos control and many of them have been applied to the Chua's circuit and to the Chua's oscillator. The aim of this Section is not to review all the approaches developed, but to give some insights on only a few important control techniques applied to the Chua's circuit. We refer to [Chen and Dong (1998)] and [Boccaletti *et al.* (2000)] for a detailed treatment on the subject of chaos control.

Both non-feedback and feedback techniques have been applied to the Chua's circuit. Among non-feedback techniques we cite the possibility of controlling the Chua's circuit by inserting in the circuit some additional components which act as a "chaotic oscillation absorber". The idea, introduced in [Kapitaniak *et al.* (1993)], is inspired to shock absorbers which can be installed in mechanical systems to suppress unpredictable dangerous vibrations. The original circuit is modified with the addition of a RLC circuit coupled with it through a resistor R_c and is governed by the following set of dimensionless equations:

$$\begin{aligned}
\dot{x} &= \alpha[y - h(x)] \\
\dot{y} &= x - y + z + \varepsilon(y_1 - y) \\
\dot{z} &= -\beta y \\
\dot{y}_1 &= \alpha_1[-\gamma_1 y_1 + z_1 + \varepsilon(y - y_1)] \\
\dot{z}_1 &= -\beta_1 y_1
\end{aligned} \tag{1.47}$$

where $\varepsilon = \frac{R}{R_c}$ is the coupling coefficient. Depending on the value of the coupling resistor different stationary states can be stabilized as discussed in [Kapitaniak *et al.* (1993)]. This technique operates with a very simple principle, without the need of feedback and control signals. However, the target behavior has to be chosen by trial and error.

As an example we consider the Chua's equations with parameters (1.9) and fix the following values for the controller parameters: $\alpha_1 = \alpha$, $\beta_1 = \beta$, $\gamma_1 = 1$ and $\varepsilon = 0.3$. The control is activated for $t > 50$ (all the simulations refer to dimensionless equations with arbitrary units). Figure 1.31 shows the trend of the x variable, demonstrating how the chaotic system is controlled to a period-1 limit cycle.

Another example of open-loop control has been already discussed in Section 1.8.1: the addition of a second sinusoidal forcing to the driven Chua's circuit may lead to stabilization of periodic orbits.

Unlike non-feedback methods which mainly use the addition of a periodic driving force or periodic parameter modulation and often imply rather

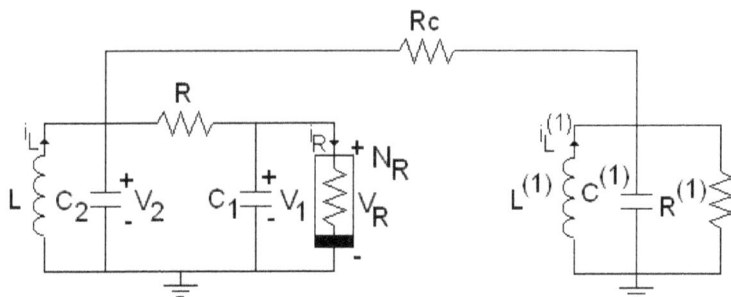

Fig. 1.30   Control of the Chua's circuit through the "chaotic oscillation absorber" [Kapitaniak *et al.* (1993)].

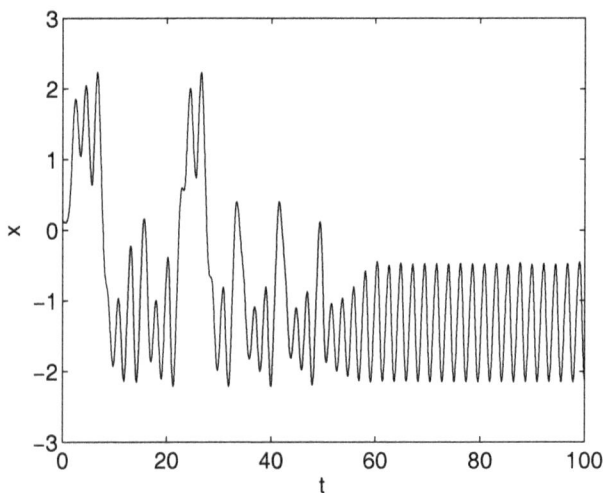

Fig. 1.31   Trend of the *x* variable of the Chua's equations controlled through the "chaotic oscillation absorber".

large modifications of the system dynamics, feedback methods exploit the properties of chaotic systems to stabilize already existing periodic orbits with small perturbations. This idea was firstly introduced by Ott, Grebogi and Yorke (OGY) [Ott *et al.* (1990)].

The peculiar properties of chaotic systems that are exploited by the OGY methods are [Boccaletti *et al.* (2000)]: the presence of infinite unstable periodic orbits embedded in the chaotic attractor; ergodicity which makes the trajectory passing close to each of such orbits; and the high sensitivity to small changes on the system current state which implies the possibility of altering the behavior of the system with small perturbations. Once determined which of the unstable periodic orbits embedded in the strange attractor is the target of the control, according to the OGY method one has to wait for the natural passage of the chaotic trajectory close to the target one. At this point, small perturbations are applied to stabilize the target orbit.

In [Johnson *et al.* (1993)] the OGY method (in particular, a one-dimensionally version of it called occasional proportional feedback) has been applied to the Chua's circuit according to the scheme shown in Fig. 1.32. The parameter to which the perturbation is applied when the trajectory deviates from the target trajectory is the negative resistance of the circuit. In parallel to this, in fact, a voltage controlled resistor is inserted, so that the overall resistance is modulated by the control signal. The experimental results obtained in [Johnson *et al.* (1993)] demonstrate that several types of periodic orbits (such as period-1, period-2 and period-4 limit cycles) can be stabilized with small control signals.

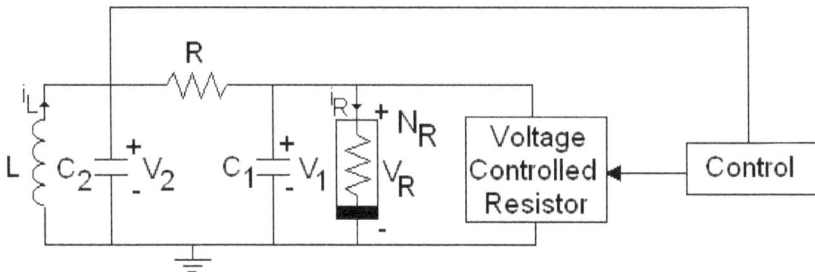

Fig. 1.32 Scheme for the occasional proportional feedback control of the Chua's circuit [Johnson *et al.* (1993)].

Among feedback methods for chaos control, one of the most effective is the use of linear feedback. This technique introduced by Chen and Dong [Chen and Dong (1993a,b,c); Chen (1993)], is based on the design of a conventional feedback controller that drives the trajectory of the system from a chaotic orbit of the system to any target trajectory of the system (for instance, one of its unstable periodic orbits).

The main result obtained by Chen and Dong may be formulated in terms of the control of the Chua's oscillator (1.42) with $k = 1$ (its generalization to $k = -1$ is however straightforward). Let $(\bar{x}, \bar{y}, \bar{z})$ be a target trajectory of system (1.42). Then, the chaotic trajectory of the Chua's oscillator can be driven to the target trajectory by adding to the equations of the Chua's oscillator the following linear control:

$$\begin{bmatrix} u_1 \\ u_2 \\ u_3 \end{bmatrix} = - \begin{bmatrix} K_{11} & 0 & 0 \\ 0 & K_{22} & 0 \\ 0 & 0 & K_{33} \end{bmatrix} \begin{bmatrix} x - \bar{x} \\ y - \bar{y} \\ z - \bar{z} \end{bmatrix} \qquad (1.48)$$

with

$$\begin{aligned} K_{11} &\geq -\alpha(m_0 - 1) \\ K_{22} &\geq 0 \\ K_{33} &\geq -\gamma \end{aligned} \qquad (1.49)$$

According to this technique, the equations of the controlled Chua's oscillator, shown in Fig. 5.11, thus become:

$$\begin{aligned} \dot{x} &= \alpha[y - h(x)] + u_1 \\ \dot{y} &= x - y + z + u_2 \\ \dot{z} &= -\beta y - \gamma z + u_3 \end{aligned} \qquad (1.50)$$

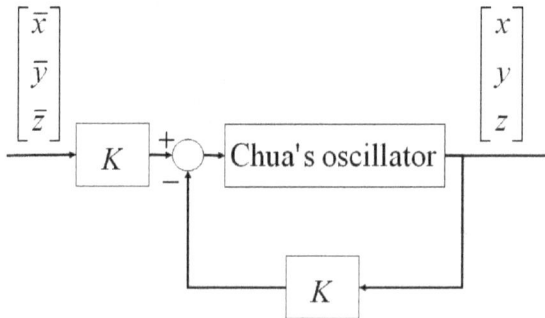

Fig. 1.33   Feedback control of the Chua's oscillator [Chen (1993)].

In the linear feedback technique the controller has a simple structure and does not need the access to the system parameters, but requires the access to many state variables.

As an example we fixed as target trajectory the equilibrium point P^+ in the Chua's circuit with parameters (1.9) and we applied the feedback controller (1.48) with $K_{11} = \alpha(m_0 - 1)$, $K_{22} = 1$ and $K_{33} = 0$ for $t \geq 50$. Fig. 1.34 shows the trend of the x variable, showing how the equilibrium point P^+ is indeed stabilized.

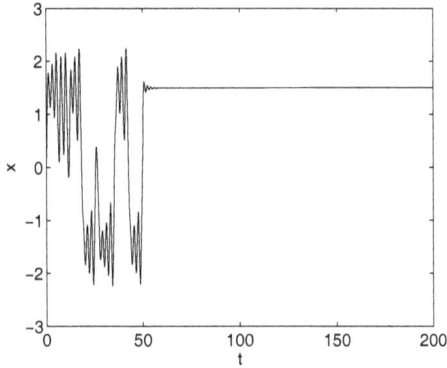

Fig. 1.34   Example of feedback control of the Chua's oscillator: stabilization of P^+.

Another closed loop method that can be applied to control the Chua's technique is the Pyragas' technique [Pyragas (1992); Pyragas and Tamasevicius (1993); Pyragas (1995, 2006)], in which a time-delayed feedback of the state variables is used to stabilize the unstable periodic orbits of the strange attractor. The technique can be suitably applied to the Chua's circuit as experimentally verified in [Celka (1994)]. According to this technique, the Chua's equations can be rewritten as follows to include the feedback term:

$$\begin{aligned}
\dot{x}(t) &= \alpha[y(t) - h(x(t))] \\
\dot{y}(t) &= x(t) - y(t) + z(t) + \varepsilon(y(t) - y(t - \tau)) \\
\dot{z}(t) &= -\beta y(t)
\end{aligned} \qquad (1.51)$$

where the feedback involves only the delayed state variable $y(t - \tau)$. Depending on the values of ε and τ different unstable periodic orbits may be stabilized. An example is shown in Fig. 1.35, where a period-2 unstable periodic orbit is stabilized ($\varepsilon = -2.1$ and $\tau = 2.2$). We notice that time-delay feedback can also be used to generate chaos in a modified Chua's circuit [Wang *et al.* (2001)].

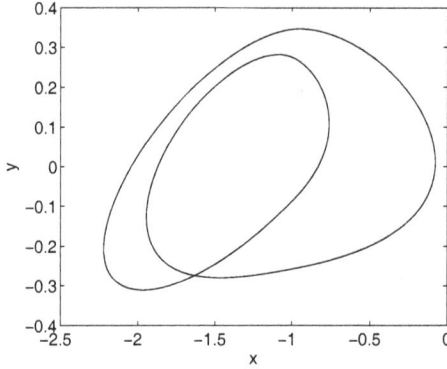

Fig. 1.35   Example of time-delay feedback control of the Chua's oscillator: stabilization of a period-2 unstable periodic orbit.

Other feedback control methods can be designed by taking into account the peculiar characteristics of the Chua's circuit. This is the case of the distortion control [Genesio and Tesi (1993)] which makes use of the fact that the structure of the Chua's circuit is that of a *Lur'e system*. Lur'e systems are described by the feedback structure shown in Fig. 1.36. As it can be easily demonstrated by direct calculation, the Chua's circuit is a Lur'e system with $w(t) = x(t)$, $n(\cdot) = f(x)$ and:

$$L(s) = \frac{\alpha(s^2+s+\beta)}{s^3+(1+\alpha)s^2+\beta s+\alpha\beta} \tag{1.52}$$

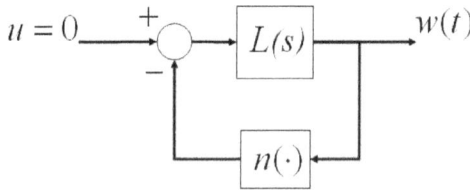

Fig. 1.36   Feedback block scheme of a Lur'e system.

Given such structure, concepts developed for the approximate analysis of Lur'e systems, and, in particular, those based on the describing function approach, can be applied to describe the birth of chaos in the Chua's circuit as the result of the interaction between an equilibrium point and a predicted limit cycle. The limit cycle is called predicted since it is derived by a first harmonic analysis which neglects higher harmonics whose amount is referred as the distortion parameter. The main idea of the approach described in [Genesio and Tesi (1993)] is that under medium levels of distortion chaos emerges, while, if distortion is controlled, the predicted limit cycle can be stabilized. The controller is, therefore, designed as a nonlinear system operating in parallel with the original nonlinearity $n(\cdot)$ and reducing the level of distortion.

Many other techniques have been developed to control chaos in the Chua's circuit. Since it is not possible to provide an exhaustive list of such approaches, we only mention some examples: control based on state space and input-output standard control techniques [Hartley and Mossayebi (1993b); Johnston and Hunt (1993)], feedback techniques [Hwang *et al.* (1996); Hwang (1997); Torres and Aguirre (1999)], return map control [Lee *et al.* (1997)], control of peak-to-peak dynamics [Piccardi and Rinaldi (2002)], impulsive control [Li *et al.* (2001)], fuzzy control [Lian *et al.* (2002)], tracking control [Puebla *et al.* (2003)] and robust tracking control through fuzzy approach [Chang (2001)], adaptive control [Ge and Wang (2000); Barone and Singh (2002)] and control based on motor maps [Arena *et al.* (2002)].

1.10   Synchronization of Chua's circuits

Synchronization of chaos is usually referred as a process wherein two (or more) chaotic systems adjust a given property of their motion to a common behavior (e.g., equal trajectories or phase locking), due to coupling or forcing [Boccaletti *et al.* (2002); Pikovsky *et al.* (2001)]. Although the topic has been extensively investigated even in relation with the Chua's circuit, we will refer here only to complete or identical synchronization, *i.e.* when two or more chaotic circuits follow the same chaotic trajectory.

Since chaotic systems exhibit high sensitivity to initial conditions and thus, even if identical and, possibly, starting from almost the same initial points, follow trajectories which rapidly become uncorrelated, appropriate techniques should be set up to obtain complete synchronization. Such techniques to couple two or more chaotic circuits can be mainly divided into two classes: drive-response (or unidirectional) coupling and bidirectional coupling. In the first case, one circuit drives another one called the response (or slave) system, while on the contrary in bidirectional coupling both the circuits are connected and each circuit influences the dynamics of the other. Both the two classes of synchronization schemes have been successfully applied to the Chua's circuit [Chua *et al.* (1993a,b)].

Let us first consider how the synchronization scheme based on drive-response coupling can be applied to the Chua's circuit. The drive-response technique was introduced by Pecora and Carroll as the first experimental proof of chaotic synchronization [Pecora and Carroll (1990)]. They consider an autonomous n-dimensional dynamical system $\dot{u} = F(u)$ and divide it into two subsystems:

$$
\begin{aligned}
\dot{v} &= G(v, w) \\
\dot{w} &= H(v, w)
\end{aligned}
\tag{1.53}
$$

where $v = (u_1, \ldots, u_m)$, $w = (u_{m+1}, \ldots, u_n)$, $G = (F_1(u), \ldots, F_m(u))$ and $H = (F_{m+1}(u), \ldots, F_n(u))$. Then, they consider a second dynamical system (identical to the first) where the variables v are sent to the subsystem $\dot{w}' = H(v', w')$ so that one obtains:

$$
\dot{w}' = H(v, w')
\tag{1.54}
$$

The two systems synchronize (*i.e.*, $w(t)$ and $w'(t)$ asymptotically have the same evolution) if all the Lyapunov exponents of the subsystem w (called conditional Lyapunov exponents) have negative real part. Such conditional Lyapunov exponents can be calculated from the Jacobian matrix $D_w H(v(t), w(t))$ of the subsystem w calculated around the given chaotic trajectory.

In the case of the Chua's circuits several system decompositions can be taken into account, not all of them lead to a subsystem with conditional Lyapunov exponents with negative real part [Chua *et al.* (1993b)].

Let us first consider the decomposition in which the x variable is used to drive the response system:

$$
\begin{aligned}
\dot{x} &= \alpha[y - h(x)] \\
\dot{y} &= x - y + z \\
\dot{z} &= -\beta y \\
\dot{y}' &= x - y' + z' \\
\dot{z}' &= -\beta y'
\end{aligned}
\tag{1.55}
$$

The implementation of this scheme is shown in Fig. 1.37(a). The conditional Lyapunov exponents of the subsystem (which in this case is linear) have negative real part: indeed, the two circuits synchronize.

Synchronization can be also achieved even if the y variable is used to drive the response system:

$$
\begin{aligned}
\dot{x} &= \alpha[y - h(x)] \\
\dot{y} &= x - y + z \\
\dot{z} &= -\beta y \\
\dot{x}' &= \alpha[y - h(x')] \\
\dot{z}' &= -\beta y
\end{aligned}
\tag{1.56}
$$

Figure 1.37(b) shows how this coupling can be implemented. Finally, we mention that when the z variable is used to drive the response system, there are conditional Lyapunov exponents with positive real part and synchronization cannot be obtained.

In the case of bidirectional coupling [Chua *et al.* (1993b)] two Chua's circuits can be coupled using one of the two simple schemes reported in Fig. 1.38. The scheme of Fig. 1.38(a) refers to coupling through the x variable. In terms of dimensionless equations it is described by the following equations:

$$
\begin{aligned}
\dot{x} &= \alpha[y - h(x)] + k_x(x' - x) \\
\dot{y} &= x - y + z \\
\dot{z} &= -\beta y \\
\dot{x}' &= \alpha[y' - h(x')] + k_x(x - x') \\
\dot{y}' &= x' - y' + z' \\
\dot{z}' &= -\beta y'
\end{aligned}
\tag{1.57}
$$

where $k_x = \frac{\alpha R}{R_c}$.

On the contrary the scheme of Fig. 1.38(b) refers to coupling through the y variable. The dimensionless equations describing this second case are the following:

$$
\begin{aligned}
\dot{x} &= \alpha[y - h(x)] \\
\dot{y} &= x - y + z + k_y(y' - y) \\
\dot{z} &= -\beta y \\
\dot{x}' &= \alpha[y' - h(x')] \\
\dot{y}' &= x' - y' + z' + k_y(y - y') \\
\dot{z}' &= -\beta y'
\end{aligned}
\tag{1.58}
$$

where $k_y = \frac{R}{R_c}$.

These two schemes have been numerically and experimentally studied in detail in [Chua *et al.* (1992, 1993b,a)]. From an experimental point of view

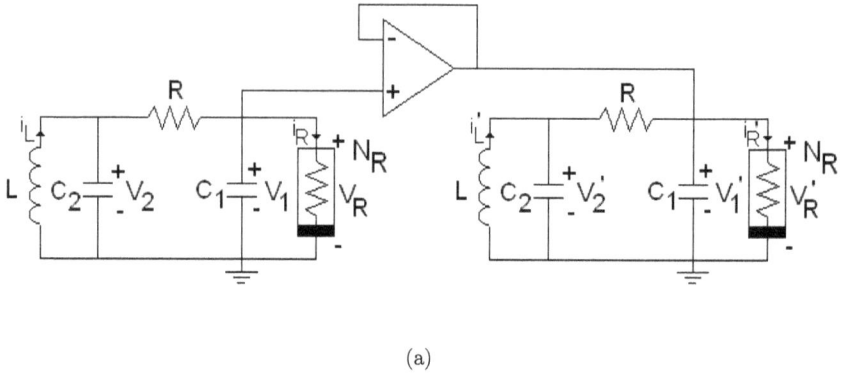

(a)

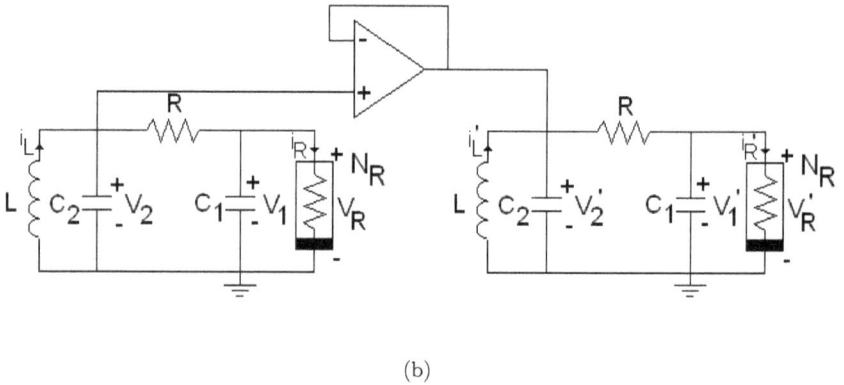

(b)

Fig. 1.37   Synchronization scheme based on drive-response coupling through the x variable (a) or the y variable (b).

[Chua *et al.* (1993a)] using a potentiometer for R_c, it can be observed that starting from a high value of such resistor the two circuits are not synchronized. When the value of the potentiometer is decreased, the two circuits do synchronize. Therefore, in both cases, synchronization is achieved for a sufficiently low value of R_c or for a sufficiently high value of the coupling parameter (k_x or k_y).

     The same synchronization technique can be applied also coupling the circuits through the z variable. In this case, the equations of the coupled Chua's circuits are:

(a)

(b)

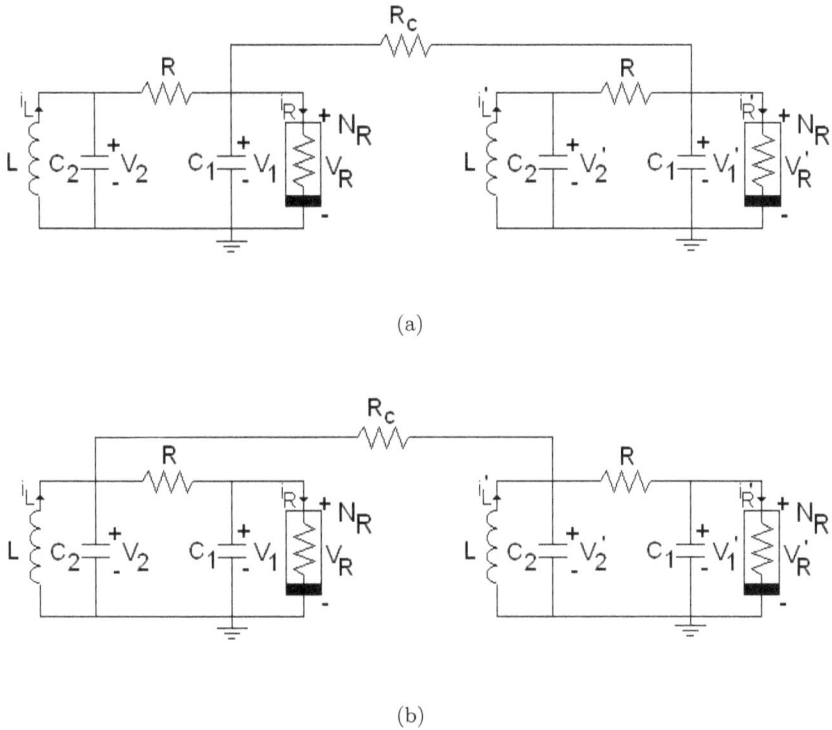

Fig. 1.38 Two suitable (and simple) synchronization schemes based on bidirectional coupling through the x variable (a) or the y variable (b).

$$\dot{x} = \alpha[y - h(x)]$$
$$\dot{y} = x - y + z$$
$$\dot{z} = -\beta y + k_z(z' - z)$$
$$\dot{x}' = \alpha[y' - h(x')]$$
$$\dot{y}' = x' - y' + z'$$
$$\dot{z}' = -\beta y' + k_z(z - z')$$

(1.59)

The implementation of such a technique is not immediate, since it requires adequate circuitry to sense the current through the inductor. However, it can be easily implemented on a Chua's circuit implementation based on state variables (such as that described in Chapter 3). From a numerical point of view, the analysis carried on in [Chua *et al.* (1993a)] gives the possibility of synchronizing two Chua's circuits through the z variable if the coupling lies within a suitable range.

The problem of chaos synchronization can be studied in a more general framework, the so called *Master Stability Function (MSF) approach* [Pecora and Carroll (1998); Boccaletti *et al.* (2006)], in which either unidirectional and bidirectional coupling (including also the case when more than two circuits are coupled) are considered.

According to the MSF approach, N identical oscillators, coupled by an arbitrary network configuration admitting an invariant synchronization manifold, are taken into account. The conditions under which such oscillators can be synchronized are unravelled by linearization of the network dynamics around the synchronization manifold.

The dynamics of each node is modelled as $\dot{\mathbf{x}}^i = \mathbf{F}(\mathbf{x}^i) - K \sum_j g_{ij} \mathbf{H}(\mathbf{x}^j)$ where $i = 1, ..., N$, $\mathbf{x}^i$ is a m-dimensional vector of dynamical variables of the i-th node, $\dot{\mathbf{x}}^i = \mathbf{F}(\mathbf{x}^i)$ represents the dynamics of each isolated node, K is the coupling strength, $\mathbf{H} : \mathbb{R}^m \rightarrow \mathbb{R}^m$ is the coupling function and $G = [g_{ij}]$ is a zero-row sum $N \times N$ matrix modelling network connections (*i.e.*, the Laplacian of the network).

According to the analysis of Pecora and Carroll [Pecora and Carroll (1998)], a block diagonalized variational equation of the form $\dot{\xi}_h = [D\mathbf{F} - K\gamma_h D\mathbf{H}]\xi_h$ represents the dynamics of the system around the synchronization manifold; where γ_h is the h-th eigenvalue of G, $h = 1, \cdots, N$. $D\mathbf{F}$ and $D\mathbf{H}$ are the Jacobian matrices of F and H computed around the synchronous state, and are the same for each block. Therefore, the blocks of the diagonalized variational equation differ from each other only for the term $K\gamma_h$. If one wants to study synchronization properties with respect to different topologies or different coupling values, the variational equation must be studied as a function of a generic (complex) eigenvalue $\alpha + i\beta$. This leads to the definition of the Master Stability Equation (MSE):

$$\dot{\zeta} = [DF - (\alpha + i\beta)DH]\zeta \qquad (1.60)$$

The maximum (conditional) Lyapunov exponent λ_{max} of the MSE is studied as a function of α and β, thus obtaining the Master Stability Function, *i.e.* $\lambda_{max} = \lambda_{max}(\alpha + i\beta)$. Then, the stability of the synchronization manifold in a given network can be evaluated by computing the eigenvalues γ_h (with $h = 2, \ldots, N$) of the matrix G and studying the sign of λ_{max} at the points $\alpha + i\beta = K\gamma_h$. If all eigenmodes with $h = 2, \ldots, N$ are stable, then the synchronous state is stable at the given coupling strength. In fact, we recall that, since G is zero-row sum, the first eigenvalue is always $\gamma_1 = 0$ and represents the variational equation of the synchronization manifold.

The MSF formalism allows to study how the overall topology of networks influences the propensity to synchronization. Specifically, it gives a necessary condition (the negativity of all Lyapunov exponents transverse to the synchronization manifold) for the stability of a complete synchronization process. With this approach, both heterogeneous and homogeneous networks, scale-free and small-world topologies, weighted and unweighted networks have been studied [Boccaletti *et al.* (2006)].

If G has real eigenvalues (for instance if it is symmetric), the MSF can be computed only as function of α. In the following we will restrict our analysis to this case. The functional dependence of λ_{max} on α gives rise to three different cases [Boccaletti *et al.* (2006)], shown in Fig. 1.39. The first case, denominated as type I, is the case in which network nodes cannot be synchronized. In the second case (type II) increasing the coupling coefficient σ always leads to a stable synchronous state. In the third case (type III), network nodes can be synchronized only if $\sigma\gamma_h$ for $h = 2, \ldots, N$ lie in the interval with negative values of λ_{max}.

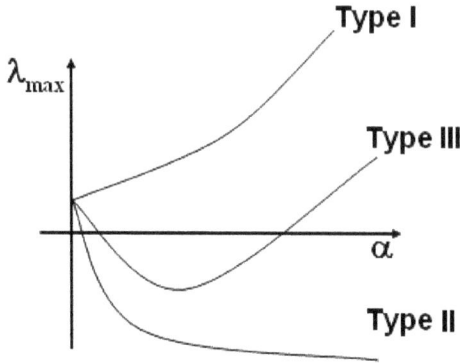

Fig. 1.39 Classification of oscillators with respect to the functional dependence of the maximum Lyapunov exponent λ_{max} on α.

As it can be noticed from equations (1.60), the MSF depends on the specific coupling function adopted. In particular, the case of bidirectional coupling through the x, y and z variables corresponds to $DH = \begin{bmatrix} 1 & 0 & 0 \\ 0 & 0 & 0 \\ 0 & 0 & 0 \end{bmatrix}$,

$$DH = \begin{bmatrix} 0 & 0 & 0 \\ 0 & 1 & 0 \\ 0 & 0 & 0 \end{bmatrix} \text{ and } DH = \begin{bmatrix} 0 & 0 & 0 \\ 0 & 0 & 0 \\ 0 & 0 & 1 \end{bmatrix}, \text{ respectively. The calculation of the}$$

MSFs corresponding to these three cases allow to conclude that both x and y coupling are type II, while the case of z coupling corresponds to a type III MSF.

From this result and from the observation that in the case of two bidirectionally coupled circuits the coupling matrix is $G = \begin{bmatrix} -1 & 1 \\ 1 & -1 \end{bmatrix}$ having eigenvalues $\gamma_1 = 0$ and $\gamma = -2$, it can be derived that, while for the x and y couplings synchronization is achieved by choosing a sufficiently large coupling coefficient, in the case of z coupling the suitable values of the coupling coefficient lies in a given interval. This conclusion perfectly agrees with the numerical and experimental results discussed in [Chua *et al.* (1993a,b)] and reported above. The advantage of the MSF analysis is that it can be extended to many other synchronization schemes.

Many further issues on synchronization of the Chua's circuit have been studied in literature. Lag synchronization [Dana and Roy (2003); Li and Liao (2004)], pulse synchronization [Fortuna *et al.* (2003b)], synchronization through impulse control [Li *et al.* (2001)], synchronization via the state observer technique [Yin and Cao (2003)], synchronization in arrays of Chua's circuits either with diffusive coupling [Belykh *et al.* (1993); Hu *et al.* (1995)] or nonlinear coupling [Sanchez *et al.* (2000)], phase synchronization [Dana *et al.* (2003); Baptista *et al.* (2003); Wang *et al.* (2005)], synchronization of Chua's circuits with time-variant parameters [Chua *et al.* (1996)] are only a limited number of examples. We refer the interested reader to the specific literature.

Synchronization can also be studied as a particular problem of chaos control [Ogorzalek (1993a,b)] in which the reference system is another circuit identical to the that controlled. In such a way the techniques discussed in Section 1.9 and based on feedback methods can be used. Further examples of approaches in which the synchronization scheme is derived applying control techniques are based on output tracking problems [Zhang and Feng (2007)] and controller based on motor maps [Arena *et al.* (2002)].

Chaotic synchronization is used in chaos-based secure communications. Mainly two different techniques, chaotic masking and chaos shift keying (or more generally parameter modulation), have been applied to use the Chua's circuit synchronization for secure communications. According to chaotic masking, the Chua's circuit is used as a generator of a chaotic signal in which the message signal is buried [Kocarev *et al.* (1992)]. In order to make possible synchronization between transmitter and receiver, the power level of the message signal should be lower than that of the chaotic signal. This makes the communication scheme sensitive to noise of the same power level of the message signal.

This drawback is avoided in the chaos shift keying technique, where the transmitted information is directly contained in the chaotic signal. In this case, in fact, the parameters of a chaotic circuit are controlled and one state variable is transmitted to the receiver consisting of a set of response subcircuits. One of such subcircuits will synchronize with the transmitter thus allowing the detection of the attractor and the associated data symbol. Such technique has been applied to the Chua's oscillator [Dedieu *et al.* (1993)] and its performance have been adequately investigated [Pinkney *et al.* (1995)]. These schemes [Dedieu *et al.* (1993); Pinkney *et al.* (1995)] are based on chaotic switching, where the message signal is digital and a set of chaotic circuit parameters represent a transmitted zero, while a second set represents a transmitted one. Chaotic parameter modulation, introduced in [Yang and Chua (1996)], is a more general scheme where one or more parameters of a Chua's oscillator are modulated and that can be also used with analog message signal.

The security of a chaos-based secure communication scheme can be increased with techniques able to make the transmitted signal more complex and to reduce the redundancy in the transmitted signal (related, for instance, to the time needed to reach synchronization). An alternative approach to the use of hyperchaos is described in [Yang *et al.* (1997)] to increase the complexity of the transmitted signal generated by a Chua's oscillator. This approach combines conventional cryptographic techniques with low-dimensional chaos. On the other hand, impulsive synchronization may be used to reduce the signal redundancy as discussed in [Yang and Chua (1997)]. Recently developed techniques [Arena *et al.* (2006); Buscarino *et al.* (2007a, 2008)] are based on the use of more than one chaotic signal (each one generated by a different chaotic circuit, including a Chua's circuit), whose linear combination is sent through a scalar channel.

Chapter 2

Implementation of the Chua's diode

2.1 Circuits based on operational amplifiers

Since several implementations of the Chua's diode are based on operational amplifiers, before illustrating them, in this Section we will briefly describe the essential properties of the operational amplifiers needed for the analysis and the design of the Chua's diode. Some of these concepts will be also used in Chapter 3 where the CNN-based implementation of the Chua's circuit is dealt with.

An operational amplifier is an electronic device with differential inputs and, typically, a single output v_o. Fig. 2.1 shows its circuital symbol and the associated power supplies. Its transfer characteristic from input to the output is nonlinear and can be expressed as follows:

$$v_o = f(v_d) = \begin{cases} -E_{sat}, & \text{if } v_d \leq -\frac{E_{sat}}{A_v} \\ A_v v_d, & \text{if } |v_d| < \frac{E_{sat}}{A_v} \\ E_{sat}, & \text{if } v_d \geq \frac{E_{sat}}{A_v} \end{cases} \tag{2.1}$$

where E_{sat} is the voltage value at which the output of the operational amplifier saturates. It depends on the internal circuitry design of the device and on the voltage supply used.

The region in which $v_o = A_v v_d$ is defined as the linear region. All the applications described in this Section refer to operational amplifiers exclusively operating in the linear region, while in the next Section the device will be used as a nonlinear component.

The operational amplifier is a versatile integrated device which performs several types of circuit operations depending on its configuration. In the ideal case, the device has high input impedance, low output impedance and a high voltage gain A_v. As a consequence of the high input impedance, no

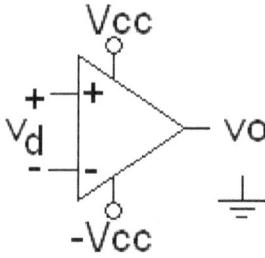

Fig. 2.1   Operational amplifier.

current flows into or out of the input terminals.

Usually the output of the operational amplifier is controlled either by negative feedback, which allows to achieve a configuration with finite voltage gain, or by positive feedback (typically in regenerative or oscillating circuits). As an example, when connected as in Fig. 2.2 the input v_{in} is related to the output v_o through the following equation:

$$v_o = -\frac{R_2}{R_1} v_{in} \qquad (2.2)$$

where the gain is fixed by the ratio between R_2 and R_1. This relationship can be derived by taking into account that the current i flowing into the resistor R_1 is given by:

$$i = \frac{v_{in} + v_d}{R_1} \qquad (2.3)$$

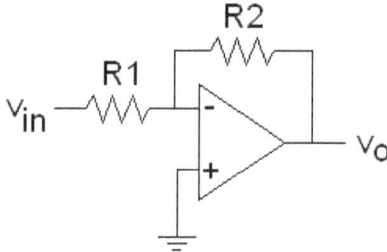

Fig. 2.2   Inverting configuration of the operational amplifier.

Since no current flows in the negative input terminal, the current in R_2 is the same in R_1, and thus:

$$v_o = -R_2 i + v_d \tag{2.4}$$

From which it can be derived that:

$$v_o = -\frac{R_2}{R_1} v_{in} - \left(\frac{R_2}{R_1} + 1\right) v_d \tag{2.5}$$

and taking into account that, in the linear region, $v_o = A_v v_d$ one obtains that:

$$\frac{v_o}{v_{in}} = -\frac{\frac{R_2}{R_1} A_v}{A_v + \left(\frac{R_2}{R_1} + 1\right)} \tag{2.6}$$

In the limit of large gain ($A_v \to \infty$), the relationship (2.2) is obtained.

Analogously, an inverting adder can be designed as shown in Fig. 2.3. In the limit of large A_v, the output is given by:

$$v_o = -\frac{R_2}{R_1} v_1 - \frac{R_2}{R_3} v_2 \tag{2.7}$$

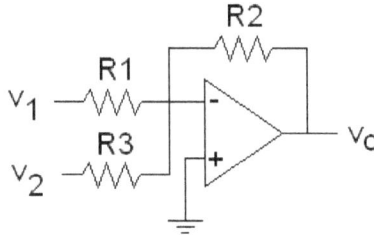

Fig. 2.3 Inverting adder configuration of the operational amplifier.

In the case of the non-inverting configuration (shown in Fig. 2.4), it can be demonstrated that the output is given by:

$$v_o = \left(\frac{R_2}{R_1} + 1\right) v_1 \tag{2.8}$$

If an algebraic adder needs to be realized, the scheme shown in Fig. 2.5 can be adopted. In particular, if the resistance R_p is chosen so that the following equality is satisfied:

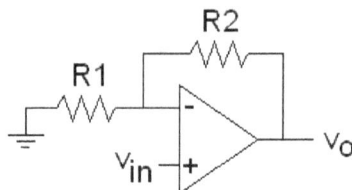

Fig. 2.4   Non-inverting configuration of the operational amplifier.

$$\frac{1}{R_1} + \frac{1}{R_2} + \frac{1}{R_f} = \frac{1}{R_3} + \frac{1}{R_4} + \frac{1}{R_p} \tag{2.9}$$

then the output of the circuit is given by:

$$v_o = \sum_i A_i v_i \tag{2.10}$$

with $A_i = \frac{R_f}{R_i}$.

Satisfying equation (2.9), in the following referred as the *gain rule*, means that the sum of the conductances at the positive input terminal of the operational amplifiers is equal to the sum of the conductances at the negative terminal. In such a way, the output depends on each single input by means of only the associated input resistor and not of the other resistors, which is very convenient from the designer perspective. We notice that, when satisfying the gain rule results in a negative value of R_p, R_f should be changed to avoid this. The algebraic adder shown in Fig. 2.5 will be used in Chapter 3.

An operational amplifier, operating in the linear region and properly configured, also allows the implementation of an integrator. The configuration is shown in Fig. 2.6, where, in the limit of large voltage gain A_v, it can be assumed that the current flowing in the resistor R is $i = \frac{v_{in}}{R}$ and is equal to that in the capacitor. Taking into account the constitutive relationship between current and voltage in a capacitor $i_c = \frac{1}{C}\frac{dv_o}{dt}$, it holds:

$$i_c = \frac{1}{C}\frac{dv_o}{dt} = -\frac{v_{in}}{R} \tag{2.11}$$

from which it can be derived that

$$v_o(t) = -\frac{1}{RC}\int_0^t v_{in}(\tau)d\tau + v_o(0) \tag{2.12}$$

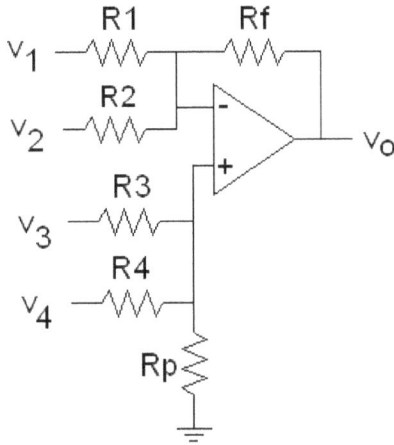

Fig. 2.5   Algebraic adder.

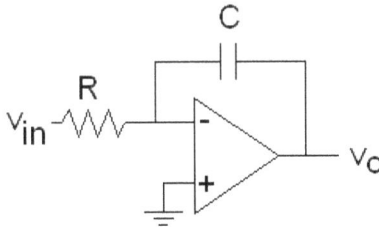

Fig. 2.6   Integrator configuration of the operational amplifier.

Finally, another configuration which exploits the properties of the operational amplifier in the linear region and that will be used in Chapter 3 is analyzed. It still performs integration of the input voltage, but in terms of filter theory the circuit is now a low-pass filter with a non-zero pole.

In the RC integrator of Fig. 2.7 the current flowing into the resistor R is:

$$i = \frac{v_o - v}{R} \qquad (2.13)$$

On the other hand, taking into account the relationship between current and voltage across a capacitor, since $i = \frac{1}{C}\frac{dv}{dt}$, one has:

$$\frac{dv}{dt} = \frac{v_o - v}{CR} \qquad (2.14)$$

and thus

$$CR\dot{v} = -v + v_o \qquad (2.15)$$

The relationship expressed by Eq. (2.15) will be used to design operational amplifier-based blocks implementing a first-order differential equation of the type $\dot{x} = -x + f(x, t)$.

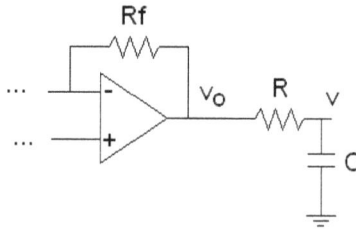

Fig. 2.7   RC integrator.

2.2   Implementation of a negative resistance

Let us first discuss the implementation of the nonlinear element in the circuit shown in Fig. 1.11. As discussed in Chapter 1, the main function of this nonlinear element is to provide a negative resistor (more precisely the nonlinear element is an *eventually passive* negative resistor). The device implementing such characteristic is shown in Fig. 2.8 [Kennedy (1992)].

Let us assume that the operational amplifier has ideal properties. Since no current flows into the positive input terminal, the current i is given by:

$$i = \frac{v - v_0}{R_1} \qquad (2.16)$$

On the other hand, since no current flows in the negative input terminal, v can be calculated from:

$$v = v_d + \frac{R_3}{R_2 + R_3} v_o \qquad (2.17)$$

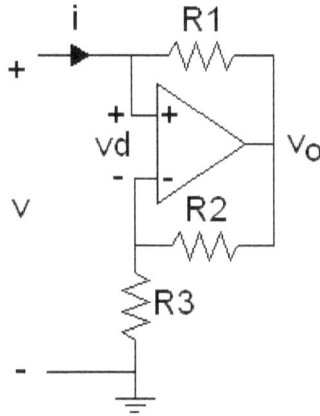

Fig. 2.8 Electronic device for the implementation of a negative resistance.

Taking into account that $v_o = A_v v_d$, Eq. (2.17) becomes:

$$v = \frac{R_2 + R_3(1+A_v)}{A(R_2 + R_3)} v_o \qquad (2.18)$$

From this relationship, v_o as function of v can be derived and then substituted in Eq. (2.16) to obtain:

$$i = \frac{(1-A_v)R_2 + R_3}{R_2 + R_3 + A_v R_3} v \qquad (2.19)$$

Under the assumption of a large open loop gain ($A_v \to \infty$) and that $R_1 = R_2$, Eq. (2.19) becomes:

$$i = -\frac{1}{R_3} v \qquad (2.20)$$

which represents the $v - i$ characteristic of the device, *i.e.*, a negative resistance.

We will now show that this device is eventually passive, *i.e.*, for large v the slope of the $i - v$ characteristic is positive. In fact, this occurs since the operational amplifier saturates. In particular, let us consider the positive saturation, *i.e.*, $v_o = E_{sat}$. Substituting this relationship in Eq. (2.16), one obtains:

$$i = \frac{v - E_{sat}}{R_1} \qquad (2.21)$$

which now represents a $v - i$ characteristic which is translated with respect to the origin and has a positive slope. The breakpoint $v = E_1$ of the whole $v - i$ characteristic of the device can be calculated by substituting $v = E_1$ in Eq. (2.17) as follows:

$$E_1 = \frac{R_2 + R_3(1 + A_v)}{A(R_2 + R_3)} E_{sat} \tag{2.22}$$

In the limit of large A_v, one obtains:

$$E_1 = \frac{R_3}{R_2 + R_3} E_{sat} \tag{2.23}$$

An analogous behavior is obtained when the operational amplifier saturates at $v_o = -E_{sat}$. The complete i_v characteristic of the device of Fig. 2.8 is thus that shown in Fig. 2.9, which is clearly the characteristic of an eventually passive negative resistor.

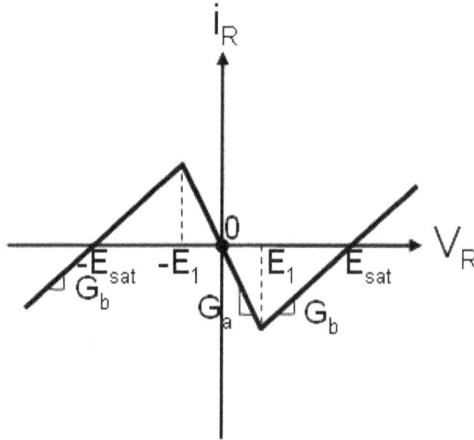

Fig. 2.9   Nonlinear characteristic of the device shown in Fig. 2.8.

2.3   Implementation of the five segment nonlinearity

The nonlinear element with the five segment characteristic appearing in the Chua's circuit can be implemented by connecting two three-segment nonlinear device of the type discussed in Section 2.2 as shown in Fig. 2.10.

Since the two devices are voltage controlled and are connected in parallel, their $v - i$ characteristic can be summed (eventually in a graphic way).

Let us indicate with $E_1 = \frac{R_3}{R_2+R_3}E_{sat}$ and $E_2 = \frac{R_6}{R_5+R_6}E_{sat}$ the break-points of the first and the second three segment device, respectively. Without any loss of generality we can assume that $E_1 < E_2$. By adding the two $v - i$ characteristics of the two branches, the five segment characteristics of the whole device can be easily obtained as shown in Fig. 2.11. The parameters of the nonlinear function are given by:

$$
\begin{aligned}
G_a &= -\frac{1}{R_3} - \frac{1}{R_6} \\
G_b &= \frac{1}{R_1} - \frac{1}{R_6} \\
G_c &= \frac{1}{R_1} + \frac{1}{R_4} \\
E_1 &= \frac{R_3}{R_2+R_3}E_{sat} \\
E_2 &= \frac{R_6}{R_5+R_6}E_{sat}
\end{aligned}
\tag{2.24}
$$

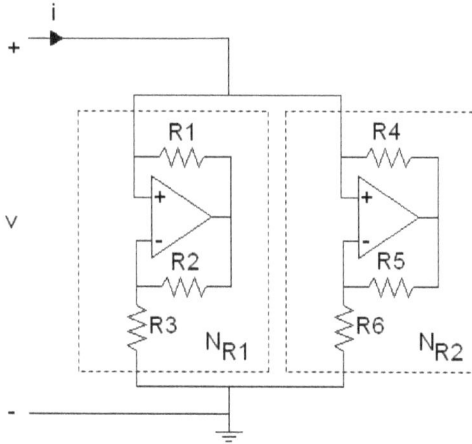

Fig. 2.10 Electronic device for the implementation of a five segment nonlinear characteristic.

2.4 Implementation of the N_R characteristic with diodes

An alternative approach to implement the five-segment characteristic is discussed in [Matsumoto *et al.* (1985)]. The scheme of such implementation is

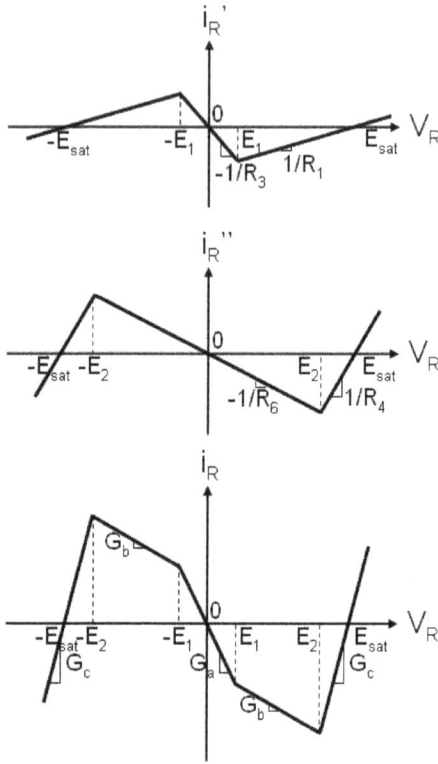

Fig. 2.11   A five segment nonlinearity can be obtained by adding two three segment piecewise linear functions.

reported in Fig. 2.12. The circuit consists of three blocks connected in parallel: one is the negative resistance block, already discussed in Section 2.2, the other two blocks, consisting of two diodes and some resistors, operate in a symmetric way. Therefore, we will focus only on one of them, *i.e.*, that containing D1. When D1 is off, it is an open circuit which disconnects node A to the rest of the circuit. Moreover, when D1 is off, at node A the potential is fixed at $\frac{R_5}{R_4+R_5}V_{cc}$. The parameters are chosen so that $\frac{R_5}{R_4+R_5} = \frac{1}{V_{cc}}$. This practically fixes the breakpoint $E_1 = 1V$. On the other hand, when D1 is on, the resistance R_5 is in parallel with the negative resistance block. Taking into account that the negative impedance block is designed so that its breakpoint $E_2 = \frac{R_3}{R_2+R_3}E_{sat} > E_1$, the whole device implements the five segment nonlinearity with the following parameters:

$$G_a = -\frac{1}{R_3}$$
$$G_b = -\frac{1}{R_3} + \frac{1}{R_5}$$
$$G_c = \frac{1}{R_1} + \frac{1}{R_5} \qquad (2.25)$$
$$E_1 = \frac{R_5}{R_4+R_5}V_{cc}$$
$$E_2 = \frac{R_3}{R_2+R_3}E_{sat}$$

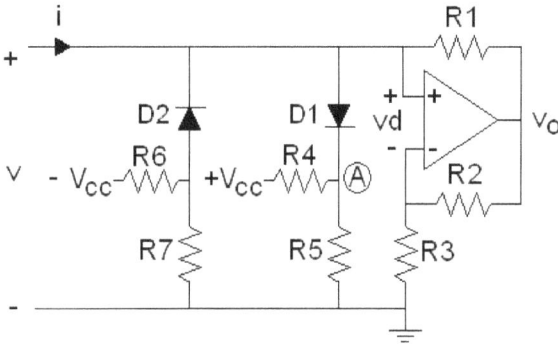

Fig. 2.12   Electronic device with diodes for the implementation of a five segment non-linear characteristic.

2.5   Two implementations of the Chua's circuit

The device shown in Fig. 2.12 and discussed in Section 2.4 constitutes a suitable implementation of the nonlinear element of the Chua's circuit (*i.e.*, the Chua's diode). In Fig. 2.13 the complete Chua's circuit based on this device is shown. The parameters are chosen according to [Matsumoto *et al.* (1985)]: $R_1 = 300\Omega$, $R_2 = 300\Omega$, $R_3 = 1.25k\Omega$, $R_4 = 46.2k\Omega$, $R_5 = 3.3k\Omega$, $R_6 = 46.2k\Omega$, $R_7 = 3.3k\Omega$, $C_1 = 5.5nF$, $C_2 = 50nF$, $L = 8.2mH$, $R = 1.33k\Omega$. In practical implementation, R is realized through a potentiometer. Figures 2.14 and 2.15 show two chaotic attractors obtained at slightly different values of the potentiometer.

In Fig. 2.16 another implementation of the Chua's circuit is shown. In this case the Chua's diode is realized through the nonlinear device of Fig. 2.10. In order to obtain a double scroll chaotic attractor, the following parameters can be used [Kennedy (1992)]: $R_1 = 220\Omega$, $R_2 = 220\Omega$, $R_3 = $

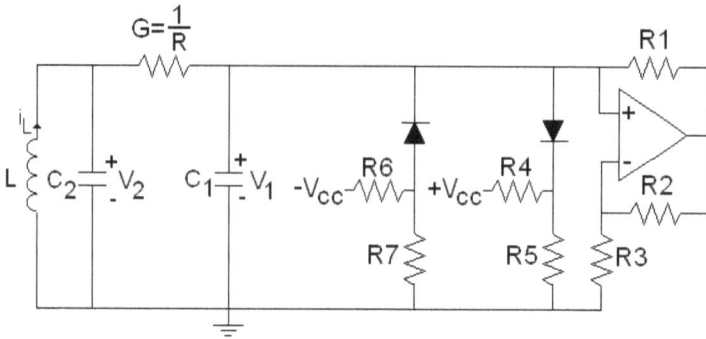

Fig. 2.13   Implementation of the Chua's circuit with the nonlinearity as in Fig. 2.12.

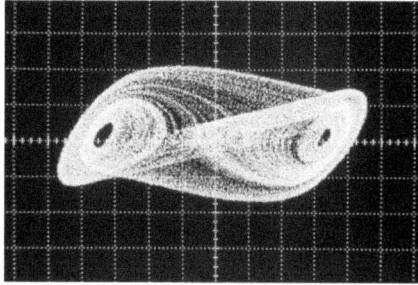

Fig. 2.14   Behavior of the Chua's circuit of Fig. 2.13 ($R = 1.30k\Omega$). Projection on the plane $v_1 - v_2$ of the double scroll strange attractor. Horizontal axis: $500mV/div$; vertical axis $200mV/div$.

$2.2k\Omega$, $R_4 = 22k\Omega$, $R_5 = 22k\Omega$, $R_6 = 3.3k\Omega$, $C_1 = 10nF$, $C_2 = 100nF$ and $L = 18mH$. The power supply for the operational amplifiers is $\pm9V$. R is realized through a potentiometer (the nominal value for the double scroll strange attractor is $R = 1.74k\Omega$).

We mention here that realizing accurate values of the inductor, especially when the further constraints of low internal resistance is added, is not trivial. In view of the realization of one of the attractors of the Chua's oscillator, this can constitute a serious difficulty. For this reason, inductorless implementations [Torres and Aguirre (2000)] of the Chua's circuit have been proposed in literature. The principle to which such implementations are based is to substitute the inductor with an equivalent circuit based on operational amplifiers. The scheme of this circuit is shown in Fig. 2.17, it

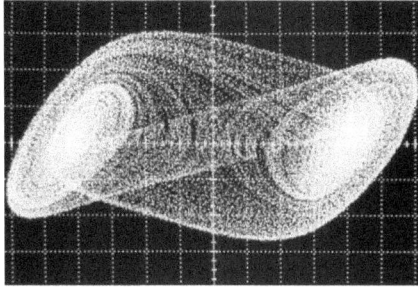

Fig. 2.15 Behavior of the Chua's circuit of Fig. 2.13 ($R = 1.31k\Omega$). Projection on the plane $v_1 - v_2$ of the double scroll strange attractor. Horizontal axis: $500mV/div$; vertical axis $200mV/div$.

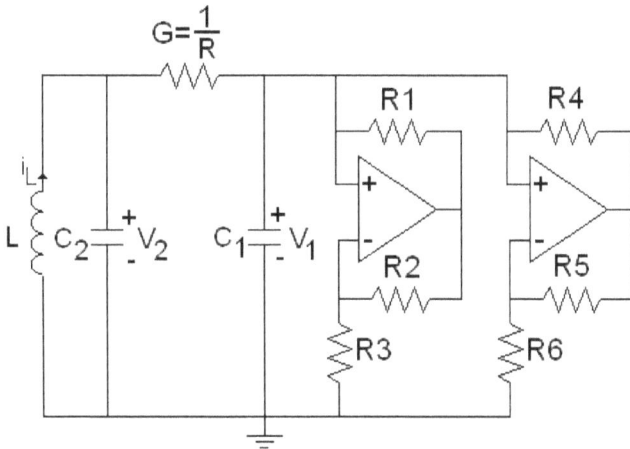

Fig. 2.16 Implementation of the Chua's circuit with the nonlinear element as in Fig. 2.10.

implements an equivalent inductor with the following value:

$$L = \frac{R_1 R_3 R_4 C}{R_2} \qquad (2.26)$$

Another implementation which avoids the use of inductors is that based on state variables and discussed in Chapter 3. These implementations have the further advantage that the third state variable is easily measurable.

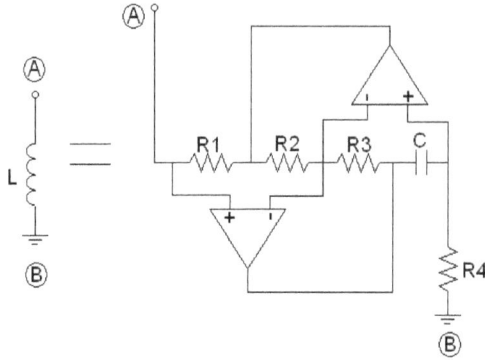

Fig. 2.17   Equivalent circuit of an inductor based on operational amplifiers.

2.6   Other implementations of the Chua's diode

Many different implementations of the Chua's diode have been reported in literature. We have described two implementations which are particularly important from the perspective of the experimentalist. Here, we cite a few other implementations which are particularly interesting.

In [Matsumoto *et al.* (1986)] it has been demonstrated that the Chua's diode can be realized using two bipolar transistors. The two transistors arranged in a proper configuration implement a piecewise nonlinearity with an inner segment with negative slope and, thus, allow to substitute the operational amplifier in the scheme of Fig. 2.13. The circuit with a two-transistor Chua's diode is shown in Fig. 2.18. Two commonly 2SC1815 bipolar transistors can be used. A double scroll attractor is reported in [Matsumoto *et al.* (1986)] for the following values of the parameters: $C_1 = 0.0053\mu F$, $C_2 = 0.047\mu F$, $L = 6.8mH$, $R = 1.21k\Omega$, $R_B = 56k\Omega$, $R_1 = 1k\Omega$, $R_2 = 3.3k\Omega$, $R_3 = 88k\Omega$, $R_4 = 39k\Omega$ and $V_{cc} = 29V$.

The Chua's diode has been also integrated into an IC chip. The first integrated implementation of the Chua's diode is discussed in [Cruz and Chua (1992)], where the internal structure is based on two operational transconductance amplifiers. The two operational transconductance amplifiers are designed in order to implement the nonlinearity of the Chua's diode without the need of other additional elements like resistances and diodes. The whole structure consists of 39 CMOS transistors. A $2\mu m$ CMOS process is adopted and the whole device occupies a chip area of 0.5mm.

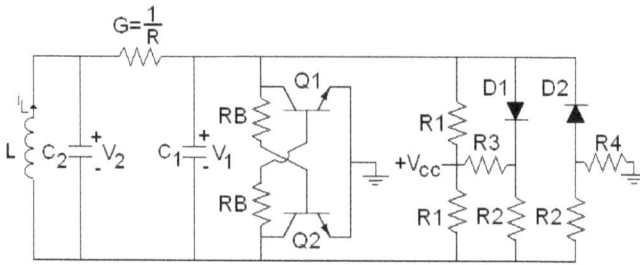

Fig. 2.18   Implementation of the Chua's diode with two transistors.

Several monolithic chip implementations of the entire Chua's circuit have been also proposed. One of them will be discussed in detail in Chapter 7. Another implementation [Cruz and Chua (1993)] extends the design of the integrated Chua's diode [Cruz and Chua (1992)] to the entire circuit. The linear resistor R is the only component not integrated in the circuit and externally implemented in order to control bifurcations of the circuit. The chip is realized using a $2\mu m$ CMOS technology and occupies a chip area of $2.5mm \times 2.8mm$. Instead, in the monolithic IC chip discussed in [Rodriguez-Vazquez and Delgado-Restituto (1993); Delgado-Restituto and Rodriguez-Vazquez (1993)] the design is based on state variables. The chip, realized in $2.4\mu m$ CMOS technology occupies a chip area of $0.35mm^2$.

As discussed in Section 1.8.3 some authors [Zhong (1994); Huang *et al.* (1996)] take into account a smooth nonlinearity for the Chua's diode and, in particular, a cubic one. Such circuit is described by the following equations:

$$\frac{dv_1}{dt} = \frac{1}{C_1}[G(v_2 - v_1) - g'(v_1)]$$
$$\frac{dv_2}{dt} = \frac{1}{C_2}[G(v_1 - v_2) + i_L] \qquad (2.27)$$
$$\frac{di_L}{dt} = -\frac{1}{L}v_2$$

with $g'(v_1) = av_1 + cv_1^3$. The implementation of such nonlinearity needs two analog multipliers. In [Zhong (1994)] the scheme of such implementation is described along with a detailed experimental study of the circuit, while the bifurcation analysis is carried on in [Khibnik *et al.* (1993b)]. A double scroll attractor appears, for instance, for the following parameters [Zhong (1994)]: $C_1 = 7nF$, $C_2 = 78nF$, $L = 18.91mH$ (with internal resistance $R_0 = 14.99\Omega$), $R = 1964\Omega$, $a = -0.59mS$ and $c = 0.02S/V^2$. Most of the results obtained with the piecewise-linear function can be extended to the circuit with the smooth nonlinearity.

In [Shi and Ran (2004)] a tunnel diode is used to implement the non-linearity of the Chua's diode. The scheme adopted is reported in Fig. 2.19, where C3 is a DC blocking capacitor and R1, R2, L1 and V_s constitute the bias circuitry of the tunnel diode. By exploiting the nonlinear characteristic of the tunnel diode, the authors report a double scroll attractors for suitable values of the parameters (given in Fig. 2.19). Compared to operational amplifier based implementation, this configuration has the advantage of expanding the bandwidth of the circuit.

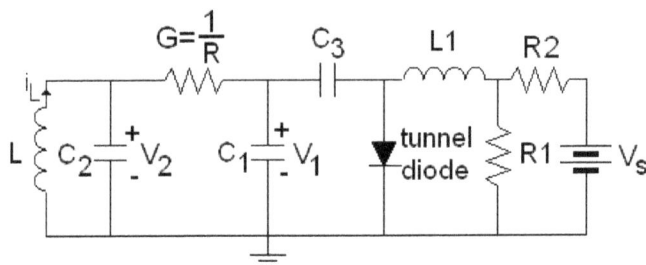

Fig. 2.19   Implementation of the Chua's diode with a tunnel diode [Shi and Ran (2004)]: $C_1 = 130pF$, $C_2 = 1300pF$, $R = 9.8\Omega$, $L_1 = 8.5nH$, $V_s = 0.45V$, $R_1 = R_2 = 5\Omega$, $L_1 = 0.1mH$, $C_3 = 100\mu F$,

Finally, we mention the interesting results obtained by [Awrejcewicz and Calvisi (2002)], where the authors propose several electromechanical implementations and a purely mechanical analog of the Chua's circuit. In their electromechanical models, electromechanical coupling is used to couple three mechanical subsystems implementing the different parts of the Chua's equations, while in the purely mechanical device only mechanical components are used. In particular, the design exploits friction to implement couplings between the different parts used.

Chapter 3

Cellular Nonlinear Networks and Chua's circuit

3.1 Brief overview on CNN architectures

Cellular Neural/Nonlinear Networks (CNNs) were introduced by L. O. Chua [Chua and Yang (1988b,a)] in 1988. His idea was to use an array of simple, identical, locally interconnected nonlinear dynamical circuits, called cells, to build large scale analog signal processing systems. The cell was defined as the nonlinear first-order circuit shown in Fig. 3.1(a), u_{ij}, y_{ij} and x_{ij} being the input, the output and the state variable of the cell respectively. The output is related to the state by the nonlinear equation:

$$y_{ij} = 0.5(|x_{ij} + 1| - |x_{ij} - 1|) \tag{3.1}$$

A CNN is defined as a two-dimensional array of MxN identical cells arranged in a rectangular grid, as depicted in Fig. 3.1(b). Each cell mutually interacts with its nearest neighbors by means of the voltage controlled current sources $I_{xy}(i,j;k,l) = A(i,j;k,l)y_{kl}$ and $I_{xu}(i,j;k,l) = B(i,j;k,l)u_{kl}$. The constant coefficients $A(i,j;k,l)$ and $B(i,j;k,l)$ are known as the *feedback* and *input cloning templates*, respectively. If they are equal for each cell, they are called *space-invariant*. If $B(i,j;k,l) = 0$ the CNN is said autonomous.

A CNN is described by the state equations of all cells:

$$C \cdot \frac{dx_{ij}}{dt} = -\frac{x_{ij}}{R_x} + \sum_{C(k,l) \in N_r(i,j)} A(i,j;k,l)y_{kl} + B(i,j;k,l)u_{kl} + I \tag{3.2}$$

with $i = 1, .2, ..., M$ and $j = 1, 2, ..., N$ where

$$N_r(i,j) = \{C(k,l) \, | \max(|k - i|, |l - j|) \leq r \}$$

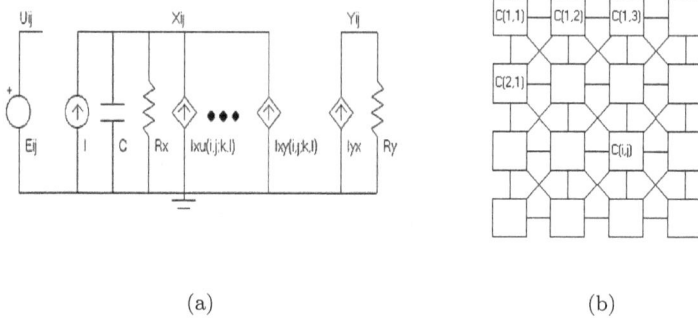

(a)   (b)

Fig. 3.1   (a) The CNN cell scheme. (b) A CNN bi-dimensional array.

with $k = 1, .2, ..., M$ and $l = 1, 2, ..., N$ is the *r-neighborhood*.

This model is known as the Chua-Yang model or linear CNN and refers to a single-layer CNN. This model can be extended to a multilayer CNN if each cell has more than one state variable.

The Chua-Yang model has been generalized in many different ways. These generalizations allow the inclusion in the model (3.2) of nonlinear interactions, direct dependence on the state variables of the neighborhood cells, different grids, and lead to a more general definition for CNNs [Chua and Roska (1993)]:

A CNN is an n-dimensional array of mainly identical dynamic systems, called cells, which satisfies two properties: (a) most interactions are local within a finite radius r, and (b) all state variables are continuous valued signals.

It follows that a more complete single-layer CNN model including some of the above mentioned generalizations is described by the following normalized state equations (setting $C = 1$, $R_x = 1$):

$$\frac{dx_{ij}}{dt} = -x_{ij}(t) + \sum_{C(k,l) \in N_r(i,j)} \left\{ \hat{A}_{ij;kl}(y_{kl}(t), y_{ij}(t)) + \right.$$
$$\left. + \hat{B}_{ij;kl}(u_{kl}(t), u_{ij}(t)) + \hat{C}_{ij;kl}(x_{kl}(t), x_{ij}(t)) \right\} + I_{ij} \qquad (3.3)$$

with

$$y_{ij} = f(x_{ij})$$

where $\hat{A}_{ij;kl}(\cdot, \cdot)$, $\hat{B}_{ij;kl}(\cdot, \cdot)$ and $\hat{C}_{ij;kl}(\cdot, \cdot)$ are two-variable nonlinear functions (the nonlinear templates) and $f(\cdot)$ is the output nonlinearity.

Significant attention has been directed to studying the dynamic properties of the various CNN models. One of the most challenging issues is surely stability [Chua and Yang (1988b)]. In fact, the particular structure, high order and nonlinearity of these systems create serious problems. Almost all kinds of dynamic behavior, ranging from simple equilibria to chaos, have been reported among the different kinds of networks.

The CNN behavior is basically dictated by the templates [Chua (1998); Chua and Roska (2005); Manganaro *et al.* (1999)]. However, the choice of templates that are suitable to achieve a desired processing task is hard to accomplish in a direct way.

This leads to the so-called *learning and design problem* [Nossek (1994); Nossek *et al.* (1993)]. The term *design* is used when the desired task can be translated into a set of local dynamic rules, while the term *learning* is used when the templates need to be obtained by learning techniques, so pairs of inputs and outputs must correspond. Some good results have been obtained with discrete-time CNNs in simple cases, but this is a really difficult problem for continuous-time models. Most of the templates currently available have been obtained by intuitive principles and refined by trial and error with the aid of simulators.

Many different spatial-temporal phenomena can be studied by using CNNs [Chua and Roska (2005); Manganaro *et al.* (1999)]. In fact, complex phenomena such as pattern generation, wave propagation, birth of spiral waves, visual processing can be investigated and reproduced with applications in different disciplines, many of them are today possible thanks to the recently developed VLSI realizations [Fortuna *et al.* (2001a)]. The emulation of these different phenomena is obtained by suitably programming the templates of the CNN, which makes it an universal paradigm for the study and the emulation of nonlinear complex dynamics. A fundamental step to achieve such general result was to show that a CNN can reproduce the complex dynamics of a chaotic circuit, and in particular of a Chua's circuit. Such implementation is based on the generalization of a CNN in terms of a State Controlled Cellular Nonlinear Network (SC-CNN) as discussed in the next Section. After briefly reviewing the SC-CNN architecture, the implementation of the Chua's circuit will be discussed.

3.2   The State Controlled Cellular Nonlinear Network

As discussed above, in the standard CNN model the state variable of each cell directly depends on the outputs and inputs of the neighbouring cells. A possible generalization of the CNN model is the so-called State Controlled CNN (SC-CNN), in which the direct dependance on the state variable is introduced. Since, in the following, only mono-dimensional arrays (instead of the more general case of two-dimensional arrays described in Section 3.1) will be used, the definition reported below is restricted to this case. In accordance with the CNN symbolism, the SC-CNN is defined as follows:

A SC-CNN is an array of nonlinear circuits $C(j)$ with the following state equation:

$$\frac{dx_j}{dt} = -x_j + \sum_{C(k) \in N(j)} \{A_{j;k}y_k + B_{j;k}u_k + C_{j;k}x_k\} + I$$
$$y_j = f(x_j)$$
$$1 \le j \le N$$
(3.4)

where:

$$f(x) = 0.5(|x+1| - |x-1|)$$
(3.5)

where x_j, y_j, u_j are the state variable, the output and the input of the cell $C(j)$, respectively, N is the number of cells, $N(j)$ is the neighbour set of the cell $C(j)$, f is the output nonlinearity function and I is the bias.

Since each cell is a first-order system, a SC-CNN made of N cells is a N-order dynamical system. By choosing the templates of the SC-CNN, *i.e.*, the connections between the cells, different dynamical systems may be reproduced. In the next Section it will be shown how a SC-CNN made of three cells (which is, thus, a third-order system) may be used to implement the Chua's equations.

3.3   CNN-based implementation of the Chua's circuit

The dynamic model of three fully connected autonomous (*i.e.* $B_j = 0 \ \forall j$), SC-CNN cells, in accordance with the state equations (3.4), is:

$$\begin{cases} \dot{x}_1 = -x_1 + a_1 y_1 + a_{12}y_2 + a_{13}y_3 + \sum_{k=1}^3 s_{1k}x_k + i_1 \\ \dot{x}_2 = -x_2 + a_{21}y_1 + a_2 y_2 + a_{23}y_3 + \sum_{k=1}^3 s_{2k}x_k + i_2 \\ \dot{x}_3 = -x_3 + a_{31}y_1 + a_{32}y_2 + a_3 y_3 + \sum_{k=1}^3 s_{3k}x_k + i_3 \end{cases}$$
(3.6)

where x_1, x_2 and x_3 are the state variables, y_1, y_2 and y_3 are the corresponding outputs.

If the parameters are fixed as follows:

$$a_{12} = a_{13} = a_2 = a_{23} = a_{32} = a_3 = a_{21} = a_{31} = 0$$
$$s_{13} = s_{31} = s_{22} = 0 \tag{3.7}$$
$$i_1 = i_2 = i_3 = 0$$

equation (3.6) becomes:

$$\begin{cases} \dot{x}_1 = -x_1 + a_1 y_1 + s_{11} x_1 + s_{12} x_2 \\ \dot{x}_2 = -x_2 + s_{21} x_1 + s_{23} x_3 \\ \dot{x}_3 = -x_3 + s_{32} x_2 + s_{33} x_3 \end{cases} \tag{3.8}$$

It can be seen how the Chua's equations are a particular case of Eqs. (3.8). In fact, assuming:

$$a_1 = \alpha(m_1 - m_0)$$
$$s_{33} = 1-$$
$$s_{21} = s_{23} = 1$$
$$s_{11} = 1 - \alpha m_1 \tag{3.9}$$
$$s_{12} = \alpha$$
$$s_{32} = -\beta$$

the Chua's equations are obtained with: $x_1 = x$, $x_2 = y$ and $x_3 = z$.

A circuit implementation of a CNN cell, inspired by [Chua and Yang (1988a)] and [Wu *et al.* (1993)], is shown in Fig. 3.2. It is constituted by three blocks:

- $B1$ forms the output nonlinearity. It exploits the natural output saturation of the amplifier $A1$; so R_7 and R_8 are chosen so that $A1$ output saturates when $|x_j| > 1$. The subsequent voltage divider (R_9 and R_{10}) is designed to scale the output voltage $-y_j$ in the range $[-1, 1]$. Hence the following design equations hold:

$$R_8/R_7 = E_{sat}/E_1$$
$$R_7/R_8 = R10/(R_9 + R_{10}) \tag{3.10}$$

where E_{sat} is the output saturation voltage of $A1$, while E_1 is its corresponding input voltage (*i.e.*, in this case $E_1 = 1V$). The input and output impedances of $B1$ and are R_7 and the parallel of R_9 and R_{10}, respectively.

- B_2 is an inverting amplifier with unity gain ($R_5 = R_6$). Its input impedance is R_5 while the output impedance is zero.
- B_3 is the core of the cell and consists of an algebraic adder in cascade with the RC integrator discussed in Chapter 2. If the parallel of the input impedance of B_1 and B_2 is very high, compared with the output impedance of block B_3 then blocks B_1 and B_2 do not load the capacitor C_j. This is clearly true if $R_7 R_5/(R_7 + R_5) \gg R_4$. In this case, the SC-CNN state equation is:

$$C_j \dot{x}_j = \frac{-x_j}{R_4} + \frac{R_3 V_1}{R_1 R_4} + \frac{R_3 V_2}{R_2 R_4} \qquad (3.11)$$

The impedances seen from the two inputs $-V_1$ and $-V_2$ are R_1 and R_2, respectively.

Let us, thus, consider three generalized cells as in [Arena *et al.* (1995a)], and represent with x_1, x_2 and x_3 each state variable and with y_1, y_2 and y_3 the corresponding outputs. If it is assumed that $V_1 = y_1$ and $V_2 = x_2$ for the first cell, $V_1 = x_1$ and $V_2 = x_3$ for the second cell and $V_1 = x_2$ and $V_2 = x_3$ for the last cell, as shown in Fig. 3.2, then the three connected cell equations are equivalent to system (3.8). Therefore they generate Chua's equations. In fact, it follows that Eqs. (3.8) are the dimensionless version of (3.11). By comparison between (3.8) and (3.11), the design of B_3 is straightforward.

3.4 Improved cell realization for SC-CNN based Chua's circuit

As shown previously, the Chua's circuit can be easily implemented by using a three cell SC-CNN, where each cell is implemented by using three operational amplifiers. In this section, a simplification of the cell circuit is discussed [Arena *et al.* (1995b)]. It starts from the following observation: the non-inverting amplifier block can be avoided if an algebraic summing amplifier, that includes the inverting operation, is introduced. Therefore, a simplified circuit realization with only two operational amplifiers for each cell can be adopted. It is shown in Fig. 3.3.

The cell circuit now consists of two blocks: the block $B1$ realizing the nonlinear function (3.5) and the block $B2$ that constitutes the cell core. The block B_1 is actually a differential amplifier stage followed by a resistive

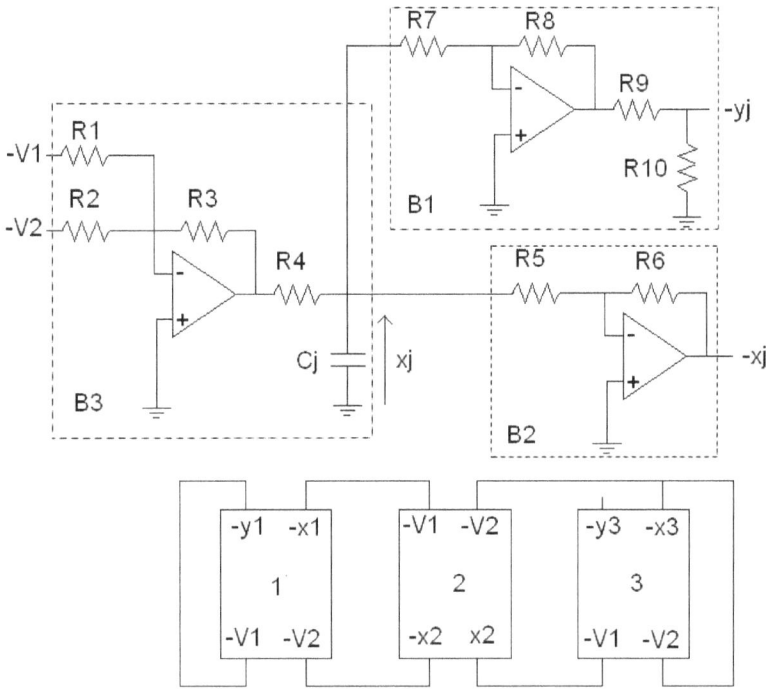

Fig. 3.2 Generalized cell circuit and connection scheme.

voltage divider. It realizes the nonlinearity of Eq. (3.5) by exploiting the natural saturation of the amplifier itself; therefore it has to be designed so that the amplifier output saturates when the input voltage reaches the desired breakpoints (*i.e.* when $|x_j| > 1$). The voltage divider attenuates the amplifier output to match the correct signal level.

The block B_2 realizes the actual cell core and is constituted by an algebraic summing amplifier stage followed by an RC integrator. If the input impedance of the block $B1$, *i.e.* $R_6 + R_{13}$, is very high compared to the output impedance of the block B_2, then the block B_1 does not sensibly influence the capacitor voltage and the following state equation holds:

$$C_j \dot{x}_j = \frac{-x_j}{R_{12}} + \frac{R_{11}V_1}{R_1 R_{12}} + \frac{R_{11}V_2}{R_2 R_{12}} - \frac{R_{11}V_3}{R_3 R_{12}} - \frac{R_{11}V_4}{R_4 R_{12}} \tag{3.12}$$

where V_1 and V_2 are the noninverting inputs and V_3 and V_4 are the

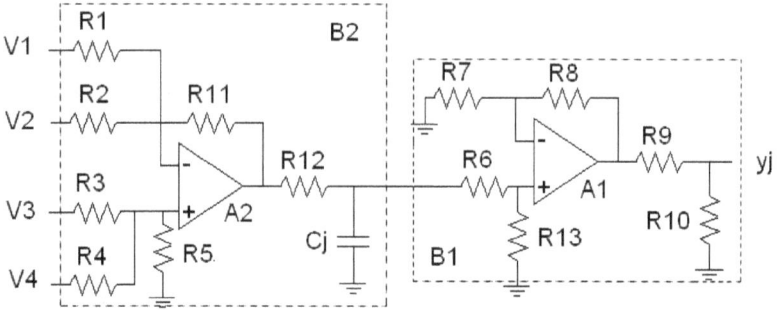

Fig. 3.3   Simplified SC-CNN cell scheme.

inverting inputs. Finally, R_5 is used to satisfy the gain rule (see Chapter 2, Eq. (2.9)) and has to be chosen as:

$$\frac{1}{R_5} = \frac{1}{R_{11}} + \frac{1}{R_1} + \frac{1}{R_2} - \frac{1}{R_3} - \frac{1}{R_4} \tag{3.13}$$

Three of these simplified SC-CNN cells are then connected together to obtain an implementation of the Chua's equations. The complete circuit is shown in Fig. 3.4. We now derive the equations governing it to show the correspondence between the circuit components of Fig. 3.4 and the parameters of the Chua's equations. From the analysis of the circuit the following equations can be derived:

$$\begin{aligned}
\dot{x}_1 &= \frac{1}{C_1 R_6}\left(\frac{R_5}{R_2}y_1 + \frac{R_5}{R_3}x_2 - x_1\right) \\
\dot{x}_2 &= \frac{1}{C_2 R_{18}}\left(\frac{R_{17}}{R_{14}}x_1 + \frac{R_{17}}{R_{15}}x_3 - x_2\right) \\
\dot{x}_3 &= \frac{1}{C_3 R_{23}}\left(-\frac{R_{21}}{R_{19}}x_2 + \frac{R_{21}}{R_{20}}x_3 - x_3\right)
\end{aligned} \tag{3.14}$$

Let us rewrite Eqs. (3.14) in a more convenient form:

$$\begin{aligned}
\dot{x}_1 &= \frac{1}{C_1 R_{18}}\frac{R_{18}}{R_6}\frac{R_5}{R_3}\left(x_2 + \frac{R_3}{R_2}y_1 - \frac{R_3}{R_5}x_1\right) \\
\dot{x}_2 &= \frac{1}{C_2 R_{18}}\left(\frac{R_{17}}{R_{14}}x_1 + \frac{R_{17}}{R_{15}}x_3 - x_2\right) \\
\dot{x}_3 &= \frac{1}{C_3 R_{23}}\left(-\frac{R_{21}}{R_{19}}x_2 + \frac{R_{21}}{R_{20}}x_3 - x_3\right)
\end{aligned} \tag{3.15}$$

Since $y_1 = \frac{R_{11}}{R_{11}+R_{12}}\frac{R_9}{R_8}(|x_1 + 1| - |x_1 - 1|)$, the nonlinearity $h(x_1)$ of the Chua's equations is implemented in the circuit as:

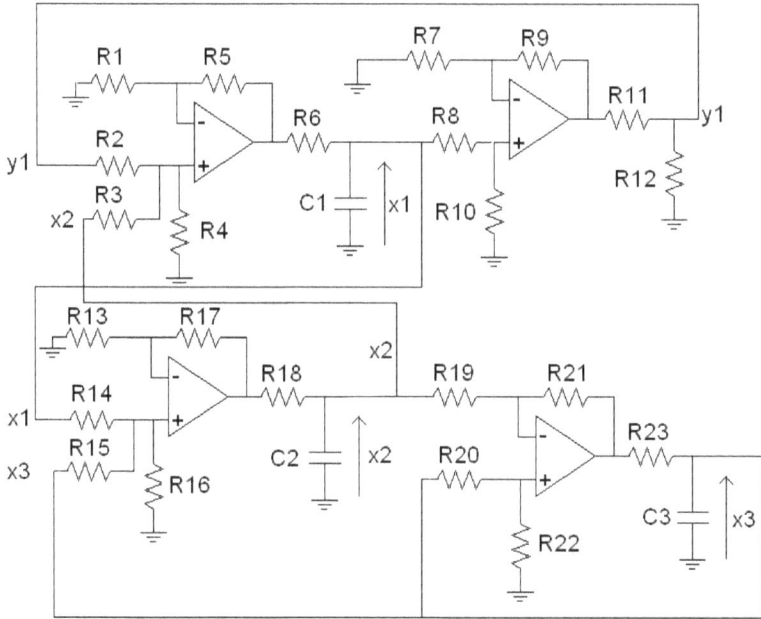

Fig. 3.4   SC-CNN Chua's circuit.

$$h(x_1) = \frac{R_3}{R_5}x_1 - \frac{R_{11}}{R_{11}+R_{12}}\frac{R_9}{R_8}\frac{R_3}{R_2}(|x_1+1|-|x_1-1|) \qquad (3.16)$$

We fix:

$$\frac{1}{C_1 R_{18}} = \frac{1}{C_2 R_{18}} = \frac{1}{C_3 R_{23}} = K \qquad (3.17)$$

as the scaling factor of the circuit. The other terms are chosen to match the Chua's equations (1.7) and to satisfy the gain rule (2.9).

The relationship between the parameters of the Chua's equations and the circuit components is, therefore, given by the following equations:

$$\begin{aligned}
\alpha &= \frac{R_5 R_{18}}{R_3 R_6} \\
\beta &= \frac{R_{21}}{R_{19}} \\
m_0 &= \frac{R_3}{R_5} - \frac{R_3 R_{21}}{R_2 R_{19}}\frac{R_{23}}{R_{22}+R_{23}} \\
m_1 &= \frac{R_3}{R_5}
\end{aligned} \qquad (3.18)$$

The SC-CNN implementation of the Chua's circuit can be easily realized with low-cost discrete components in a laboratory equipped with a power supply and an oscilloscope. A TL084 which contains four operational amplifier may be used. The power supply can be fixed to $\pm 15V$. The following values for the circuit components have been used: $R_1 = 4k\Omega$, $R_2 = 13.3k\Omega$, $R_3 = 5.6k\Omega$, $R_4 = 20k\Omega$, $R_5 = 20k\Omega$, $R_6 = 380\Omega$ (potentiometer), $R_7 = 75k\Omega$, $R_8 = 75k\Omega$, $R_9 = 1M\Omega$, $R_{10} = 100k\Omega$, $R_{11} = 12.1k\Omega$, $R_{12} = 1k\Omega$, $R_{13} = 51.1k\Omega$, $R_{14} = 100k\Omega$, $R_{15} = 100k\Omega$, $R_{16} = 100k\Omega$, $R_{17} = 100k\Omega$, $R_{18} = 1k\Omega$, $R_{19} = 8.2k\Omega$, $R_{20} = 100k\Omega$, $R_{21} = 100k\Omega$, $R_{22} = 7.8k\Omega$, $R_{23} = 1k\Omega$, $C_1 = C_2 = C_3 = 100nF$.

Although the power supply in our circuit has been fixed to $\pm 15V$, it can be easily scaled. This requires to change accordingly the components of the block implementing the nonlinearity, which has to be designed taking into account the new value of the operational amplifier saturation E_{sat}. In such a way the SC-CNN based Chua's circuit can be powered by voltage batteries and easily realized without any particular electronic instrument. Indeed, the waveforms generated by the circuit can be visualized into a PC using one of the free available softwares for the emulation of a virtual oscilloscope.

Some experimental results generated by the SC-CNN circuit have been already shown in the previous Chapters. In particular, the circuit has been used to generate the double scroll strange attractor shown in Fig. 1.3. The circuit has been also used to illustrate the sequence of period-doubling bifurcations shown in Fig. 1.17, the period-3 limit cycle of Fig. 1.18 and the single scroll attractor of Fig. 1.17(f). The parameter varied in such experiments is the resistor R_6.

3.5   SC-CNN-based implementation of the Chua's oscillator

The SC-CNN approach can be also used to implement the equations of the Chua's oscillator (1.42). Only slight changes of the circuit in Fig. 3.4 are required, or, in terms of CNN, only the reprogramming of the templates is needed. More in details, as an example we refer to the attractor shown in Fig. 1.23(c), which can be described by Eqs. (1.42) with the following parameters:

$$m_0 = 0.7562$$
$$m_1 = 0.9575$$
$$\alpha = -1.5601$$
$$\beta = 0.0156 \tag{3.19}$$
$$\gamma = 0.1581$$
$$k = -1$$

Following the SC-CNN approach described above, the circuit shown in Fig. 3.5 is obtained. With respect to the SC-CNN Chua's circuit the following changes have been taken into account:

- Since $k = -1$, the signs of the terms appearing in the second equation of the circuit dynamics are reversed. For this reason, R_{14} and R_{15} are now connected to the negative input terminal of the operational amplifier implementing the RC integrator of the variable x_2.
- Analogously, R_{19} is now connected to the positive input terminal, since $-k\beta y = |\beta|y$. Furthermore, since β is quite small, it is realized using two operational amplifiers in cascade. So, a further operational amplifier is included in the circuit of Fig. 3.5.
- In the block implementing the RC integrator of the variable x_3 the further term $-\gamma z$ appearing in the third equations of (1.42) is implemented by choosing a value of $R_{21}/R_{20} \neq 1$.

It should be noticed that, although m_0 and m_1 are now both positive, the block implementing the nonlinearity does not require any change, since using equations (3.18) the values of the circuit components can be also fixed to obtain $m_0 > 0$. The relationships between the parameters of the Chua's oscillator and the circuit components can be therefore summarized as follows:

$$\alpha = \frac{R_5 R_{18}}{R_3 R_6}$$
$$\beta = \frac{R_{21}}{R_{19}} \frac{R_{28}}{R_{25}}$$
$$\gamma = \frac{R_{21}}{R_{20}} \tag{3.20}$$
$$m_0 = \frac{R_3}{R_5} - \frac{R_3 R_{21}}{R_2 R_{19}} \frac{R_{23}}{R_{22}+R_{23}}$$
$$m_1 = \frac{R_3}{R_5}$$

Figure 3.6 shows the projection of the attractor obtained from the circuit, once fixed the parameters to match (3.19) as follows: $R_1 = 22k\Omega$, $R_2 = 112k\Omega$, $R_3 = 22k\Omega$, $R_1 = 23.3k\Omega$, $R_5 = 23k\Omega$, $R_6 = 670\Omega$,

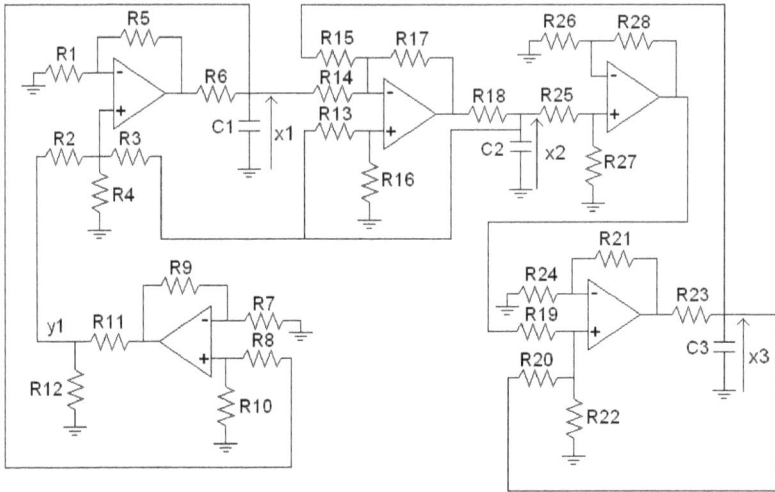

Fig. 3.5   SC-CNN-based implementation of the Chua's oscillator.

$R_7 = 75k\Omega$, $R_8 = 75k\Omega$, $R_9 = 1M\Omega$, $R_{10} = 1M\Omega$, $R_{11} = 12.1k\Omega$, $R_{12} = 1k\Omega$, $R_{13} = 50k\Omega$, $R_{14} = 100k\Omega$, $R_{15} = 100k\Omega$, $R_{16} = 100k\Omega$, $R_{17} = 100k\Omega$, $R_{18} = 1k\Omega$, $R_{19} = 630k\Omega$, $R_{20} = 84.6k\Omega$, $R_{21} = 100k\Omega$, $R_{22} = 23.5k\Omega$, $R_{23} = 1k\Omega$, $R_{24} = 22k\Omega$, $R_{25} = 500k\Omega$, $R_{26} = 22k\Omega$, $R_{27} = 15.5k\Omega$, $R_{28} = 50k\Omega$, $C_1 = C_2 = C_3 = 100nF$.

As it can be noticed the chaotic attractor generated from the circuit matches that of the Chua's oscillator shown in Fig. 1.23(c).

As shown in this Chapter, using a CNN either a Chua's circuit or a Chua's oscillator can be implemented. This implementation is at the basis of many of the circuits described in the following Chapters. For instance, the programmable Chua's circuit of Chapter 6, the integrated chip designed and tested in our laboratory at the University of Catania and described in Chapter 7 and the organic Chua's circuit discussed in Chapter 9 are all based on the CNN implementation introduced here.

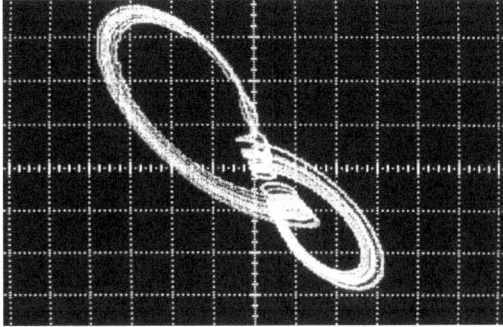

Fig. 3.6   Projection on the plane $x_2 - x_3$ of the chaotic attractor generated from the SC-CNN Chua's oscillator. Horizontal axis: $2V/div$; vertical axis $200mV/div$.

Chapter 4

Frequency switched Chua's circuit: experimental dynamics characterization

4.1 Model

In this chapter we discuss the so called *frequency switched Chua's circuit* [Cantelli *et al.* (2001)], consisting of a traditional Chua's circuit, in a chaotic double scroll parameter configuration, which is enhanced in order to generate different attractors by varying the frequency of an external control signal.

Threshold circuit generating chaos are not new in literature [Tang *et al.* (1983)]. Compared to other studies in which the intermittency between chaotic phenomena has focused on Chaos-Chaos Intermittency [Anishchenko *et al.* (2003)], the aims and conclusions of our analysis are different. In the following we examine the behavior of the system with respect to the frequency of the control signal showing that variations on this parameter alone are sufficient to change substantially the circuit dynamics.

In this Chapter the results of both numerical simulations and experimental studies are illustrated. Schematic diagrams of the behavior of the system versus the frequency of the applied control signal are sketched out. By comparing the diagram obtained for increasing frequencies and the one obtained for decreasing frequencies, hysteresis phenomena appear. This behavior, not previous discovered in other circuits, allows us to conclude that the system, for a fixed frequency value, has two different chaotic attractors corresponding to different initial conditions. This has been verified also by performing numerical simulations on the system equations.

As shown in Chapter 3, the Chua's circuit can be implemented by using three cells of a State-Controlled Cellular Neural Network. We start from this implementation and introduce a simple modification in order to set up a threshold in the Chua's circuit, that allows us to switch continuously from a two cell scheme to the classical three cell configuration, and vice versa. This purpose has been achieved by inserting an nMOS device in the SC-CNN Chua's circuit, as shown in Fig. 4.1.

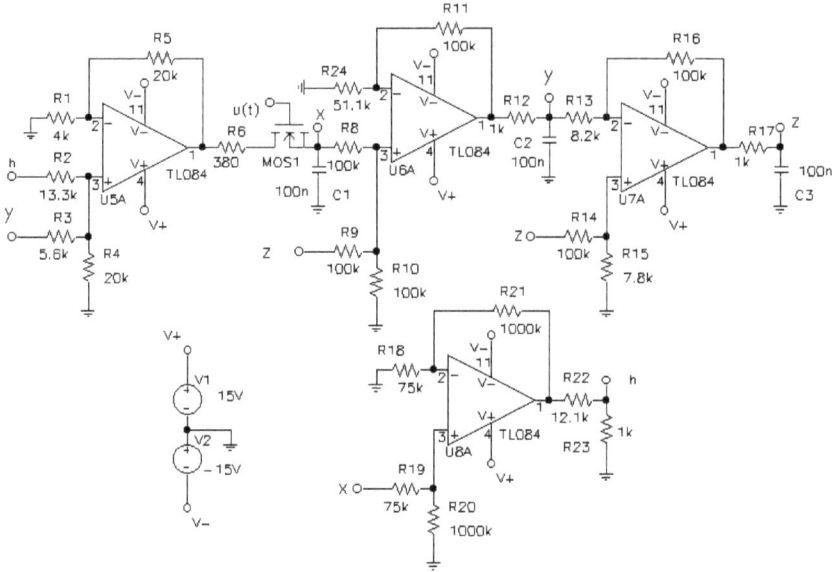

Fig. 4.1   Scheme of the frequency switched Chua's circuit.

Taking into account the circuit modification introduced above, the Chua's equations (1.7) can be written as follows:

$$
\begin{aligned}
\frac{dx}{dt} &= u(t) \cdot \alpha[y - h(x)] + [1 - u(t)] \cdot \left(-\frac{x}{\tau}\right) \\
\frac{dy}{dt} &= x - y + z \\
\frac{dz}{dt} &= -\beta \cdot y \\
h(x) &= m_1 \cdot x + 0.5 \cdot (m_0 - m_1) \cdot [|x + 1| - |x - 1|]
\end{aligned}
\tag{4.1}
$$

The introduced model differs from that described by Chua's equations in the presence of $u(t)$ and τ. The latter parameter represents the normalized time constant that regulates the discharge of the capacitor associated with the variable x when the switch is off, $u(t)$ has been introduced in order to model the effects of the nMOS. This device is driven in order to operate in nonlinear regions. It can be either "off" or "on", depending on the value of the driving signal $u(t)$. For this reasons in system (4.1) it can be assumed that $u(t)$ alternates between the discrete values 0 and 1. The values of the parameters in (4.1) are chosen as in Section 1.5.1 to obtain the double scroll behavior of the original system, while the value of τ is assumed equal to 150.

It is evident that for $u(t) = 1$ Eqs. (4.1) are identical to the Chua's equations, while, when $u(t) = 0$ the first equation of (4.1) is uncoupled from the last two and any chaotic dynamics cannot be obtained. Focusing on the circuit, when the nMOS device is closed the circuit is exactly a Chua's circuit, while, when the nMOS is open, the capacitor that integrates the first equation of (4.1) is isolated by its effective current input and consequently discharges.

4.2   Experimental results

The experiments reported in this Chapter have been performed under the assumption that $u(t)$ is a square wave signal with frequency f_u. In the following, experimental results about the behavior of the circuit with respect to the frequency f_u are discussed.

The frequency f_u of the square wave $u(t)$ has been varied from 250Hz to 6kHz, which is in the most significative frequency spectrum of the above circuit. The behavior of the system is different when operating in either low frequencies (250Hz-1kHz) or higher frequencies (1kHz-6kHz).

In the first case we observe that if the frequency of $u(t)$ is low, then the double scroll strange attractor survives, but, if this frequency is increased, then only a single band attractor is visible. When f_u varies in this region, it may be possible to observe either a chaotic attractor, or a periodic orbit. In Fig. 4.2 some examples of the trajectories on the $x - y$ phase plane are reported, together with their respective spectra for both the original circuit and the frequency switched one.

Moreover, two different experiments leading to different results are considered. In the first experiment the frequency f_u has been increased by starting by a value of 250Hz up to 1kHz. In the second experiment, the frequency f_u has been decreased back from 1kHz to 250Hz. We have noticed that the behavior of the circuit is not the same in the two experiments, thus revealing a hysteresis behavior. This result is emphasised in Fig. 4.3, where the behavior of the circuit versus frequency has been sketched.

Figure 4.3(a) deals with the first experiment (increasing frequency). In this picture, three different regimes of the circuit are schematically depicted

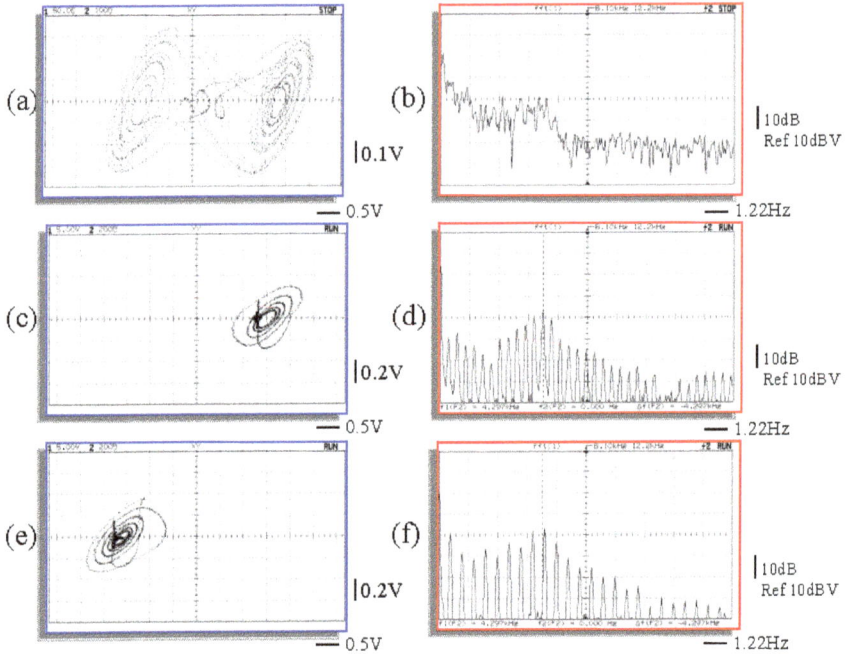

Fig. 4.2   (a) Double Scroll of the original Chua's circuit. (b) Fast Fourier Transform (FFT) of x as in (a). (c) Phase plane $x - y$ for $f = 488$ Hz. (d) FFT(x) as in (c). (e) Phase plane $x - y$ for $f = 356$ Hz. (f) FFT(x) as in (e).

versus the frequency f_u. Regime "A" indicates that the state trajectory is bounded in the $x > 0$ region, regime "C" refers to an attractor bounded in the $x < 0$ region, and regime "B" indicates that the state trajectory is not bounded in any of these regions. It has to be remarked that through this analysis we investigate only on the region of the phase plane in which the attractor is confined, but not on the presence of chaos. In fact, regardless of the region in which the dynamics evolves, both chaotic and periodic behavior can be observed. Figure 4.3(b), obtained for decreasing frequencies, does not reveal regime "A". By comparing these results the hysteresis phenomenon emerges, as illustrated in Fig. 4.3(c).

Therefore, for a given frequency f_u of the control signal, the circuit presents two symmetrical different attractors, which should take place in correspondence to two different sets of initial conditions. In order to verify this hypothesis we carried out some numerical simulations on system (4.1), by fixing the values of the frequency for which the hysteresis has

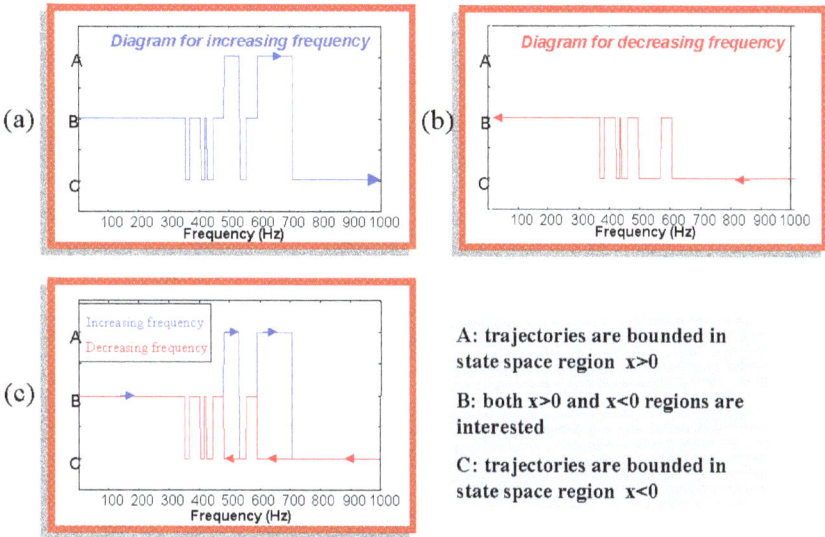

Fig. 4.3   (a) Behavior of the circuit for increasing frequency. (b) Behavior of the system for decreasing frequency. (c) Superimposition of the two cases, emphasizing hysteresis.

been revealed on circuit, and verifying that starting from different initial conditions, two different attractors emerged. Our simulations confirmed the results obtained experimentally. Moreover, while when we performed real experiments on the circuit, the long exploration of the state space necessary to understand the behavior of system (4.1) with respect to different initial conditions can be avoided, when numerical simulations are carried out, this step is mandatory. Therefore, the experimental case offers a immediate method to support the study of the existence of multiple attractors.

The results of the numerical simulations are reported in Fig. 4.4. We have chosen (refer to the dimensionless system equation (4.1)) a period of $T = 20$ for the applied $u(t)$. Obviously, the two symmetrical single scrolls shown in Fig. 4.4 appear for symmetrical initial conditions $(-0.1\ 0\ 0)$ in Fig. 4.4(a) and $(0.1\ 0\ 0)$ in Fig. 4.4(b).

As highlighted above, in Fig. 4.3 it is generically indicated for which values of f_u the trajectories are bounded in one of the two regions $x > 0$ and $x < 0$. No specific informations are available to understand for which values the system behaves chaotically or not. In effect the behavior of the system cannot be schematised in such a simplified way. In both the regions the birth of chaos manifests itself in the same way. An example

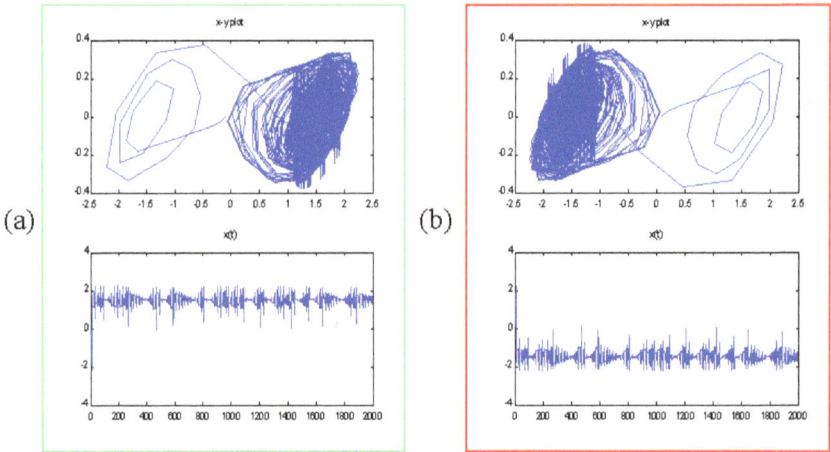

Fig. 4.4 $x - y$ plane (up) and $x(t)$ (down) for two different (symmetrical) initial conditions.

of this evolution is illustrated with the help of Fig. 4.5(a) which shows that for $f_u = 700Hz$, when chaos is triggered by the switch on, the system dynamics starts always from the same initial conditions, thus resulting in a cyclic orbit. Moreover, for a higher value of f_u ($f_u = 786Hz$, in Fig. 4.5(b)) the same phenomenon does not occur. Alternatively, the trajectory ends at a point A or at a point B. So the system after the 'resetting' of the nMOS transistor alternatively starts from two different initial conditions (A and B) thus resulting in a period-2 limit cycle.

If the value f_u is increased further, the number of these initial points becomes higher, therefore it is possible to observe a strange attractor, which is single band.

We remark that in the case in which the trajectory is a limit cycle (for example in Fig. 4.5(c)) the spectra analysis shows significant components at $f = f_u$, $f = 2f_u$, $f = 3f_u$, ... where f_u is the frequency of the signal $u(t)$. Another interesting observation that can be drawn starting from the spectra analysis is that the maximum amplitude in the spectra is always located in a well defined area around 4kHz.

A picture of the experimental global behavior is shown in Fig. 4.6. The trends of x for different values of f_u have been sampled and stored. Then, Fast Fourier Transforms (FFT) of these signals are performed. These are also functions of f_u. In such a way a 3-dimensional picture can be

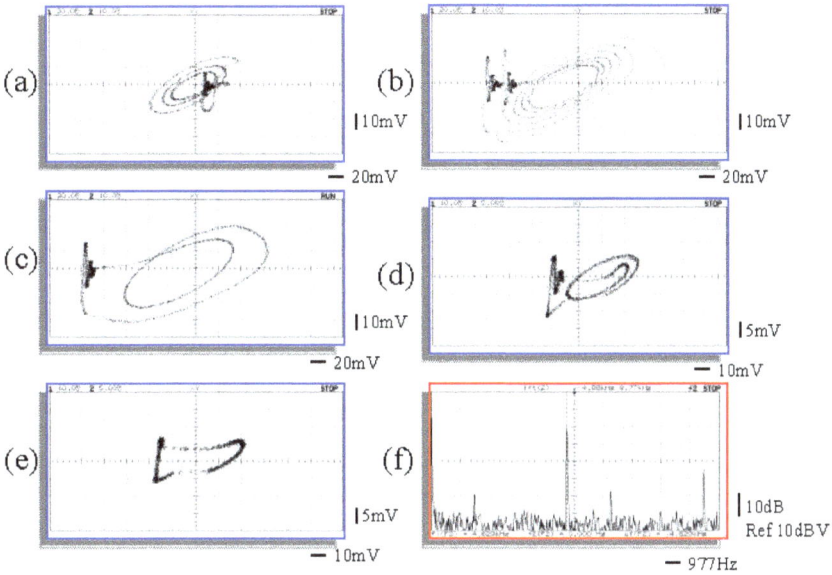

Fig. 4.5   Dynamic behavior of the system for different values of f_u: phase plane $x - y$
(a) $f_u = 700Hz$ (b) $f_u = 786Hz$ (c) $f_u = 1146Hz$ (d) $f_u = 1370Hz$ (e) $f_u = 3250Hz$
(f) FFT(x) in the case (e).

constructed, in which the various FFTs of variable x for different value of
the frequency f_u are reported. This is what is shown in Fig. 4.6, obtained
for decreasing values of f_u. In Fig. 4.6(a) it is possible to observe for
which values of f_u the system is characterized by the double scroll strange
attractor, or not, in fact the DC component is zero in the former case.
However, information about the right or left scroll is lost. By removing the
DC component, as in Fig. 4.6(b), it is possible to observe when the system
behaves chaotically or not, because chaos is evidently characterized by a
broad-band spectrum, whereas periodic regime is revealed by an almost
peak-like spectrum. Moreover, the white narrow bands displayed in a ray-
like fashion on the $f - f_u$ plane, constitute an experimental proof of the
relationship between the frequency of the control signal and that of the
oscillations of the circuit ($f = f_u$, $f = 2f_u$, $f = 3f_u$, ...).

Nevertheless, the FFT of the signal has been further performed. As
chaotic signals are characterized by a noise-like spectrum, opposite to periodic ones that are characterized by a peak-like discrete spectrum, the

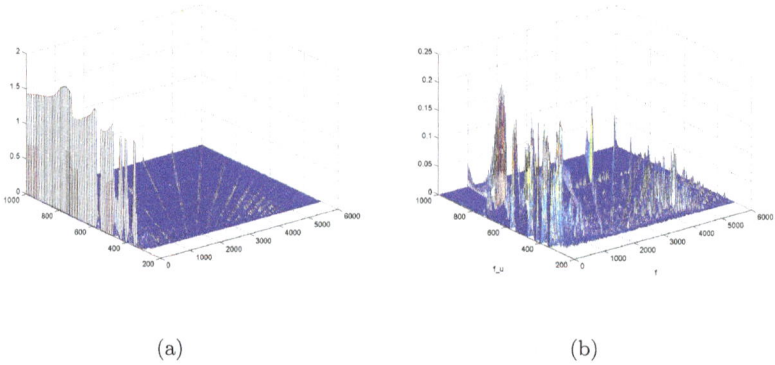

(a)   (b)

Fig. 4.6   (a) FFT of the variable x with respect to the frequency f_u. (b) FFT of the variable x with respect to the frequency f_u, where the DC component has been taken off.

standard deviation of the FFT can be used as a measure of the dispersion of the band of the signal. Therefore, the standard deviation must be high when the system is chaotic; this is shown in Fig. 4.7(b). In Fig. 4.7(b) the frequency f_M at which the FFT assumes its maximum value is shown in order to reveal the main oscillation frequency, which is very significant in the case of periodic oscillations (*i.e.* when the variance is low). It is worth noting that f_M is always located in a region around 4kHz. Moreover, the tracts where f_M is high are characterized by trajectories in a region "A" or "C", and in these tracts f_M grows linearly with the frequency of the applied $u(t)$, as already noticed before.

Further investigations on the circuit behavior are performed at higher frequencies than the ones examined above. In Fig. 4.5(c, d, e, f) several examples are reported. The principal difference of this analysis from the one at lower frequencies is twofold. Firstly, it can be noted that in this case the oscillations the system exhibits have smaller amplitudes than those obtained for low frequencies (in Fig. 4.5(d) and Fig. 4.5(e) different scales are evident). Moreover, at higher frequencies the circuit is not chaotic. The reason can rely on what was discussed before: the first component in the spectra is related to the frequency of the applied signal; if the first component is too high, a chaotic behavior is not admitted. Finally, Fig. 4.5(f) shows the spectrum when a signal $u(t)$ of frequency $f = 3250Hz$ is applied; in this case oscillations are very regular, as also verified by numerical

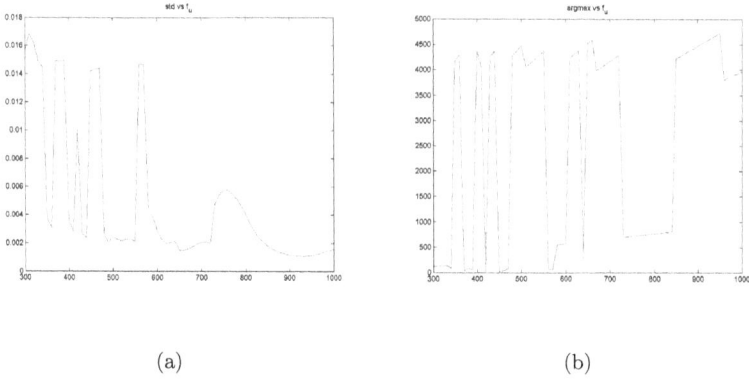

(a)   (b)

Fig. 4.7   (a) Standard deviation of $\mathrm{FFT}(x)$ versus the frequency f_u of the applied signal. (b) Frequency f_M at which $\mathrm{FFT}(x)$ assumes its maximum value versus f_u.

simulations.

In conclusion, in this Chapter it is shown how the introduction of a switching device controlled by an external signal can further enrich the dynamics of the Chua's circuit.

Chapter 5

Photo-controlled Chua's circuit

In this Chapter a Chua's circuit based on a photoresistor nonlinear device [Rahma *et al.* (2009)] is introduced and the effects of controlling it by a light source are experimentally investigated. Light control affects the dynamics of the circuit in several ways, and the circuit can be controlled to exhibit periodicity, period-doubling bifurcations and chaotic attractors. The dynamics of the circuit that operates at frequencies up to kilohertz is strongly influenced by using periodic driving signals at low frequencies. In particular, experimental results show that an unstable intermittent behavior can be observed and that this can be stabilized by using feedback.

5.1 The nonlinear photo-controlled device

Let us consider the attractor shown in Fig. 5.1. It represents the input-output behavior of the nonlinear circuit shown in Fig. 5.2 where the input signal is a chaotic signal (more precisely, it is a chaotic signal obtained by adding the x variable of the classical Chua's circuit and a DC offset). The V_o signal is the output of the nonlinear device, which, even if it has simple structure and function, allows us to introduce a new version of the Chua's circuit with interesting properties. First of all, the characterization of the circuit shown in Fig. 5.2 is given.

The device consists of a photoresistor and a light source (*i.e.*, a LED). The two components are placed in a dark small tube which isolate them from external light conditions. A fixed distance of $d = 1cm$ is considered between the LED and the photoresistor.

The behavior of the device strongly depends on the voltage V_i applied to the LED. The circuit introduces a low-pass filtering effect which depends on the offset voltage of the input signal as shown in the device characterization

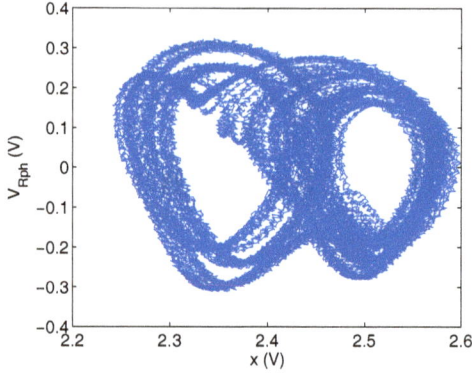

Fig. 5.1   Attractor obtained by considering a chaotic signal plus a constant voltage offset $V_{DC} = 2.6V$ as input of the photoresistor-based device.

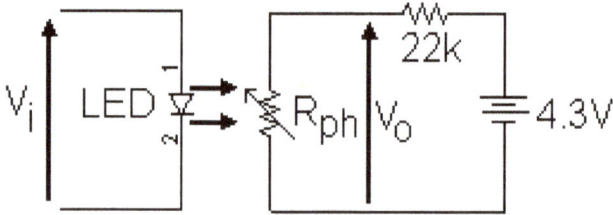

Fig. 5.2   The nonlinear device made of a photoresistor controlled by an input signal through a LED.

given in Fig. 5.3. The frequency response of the device was obtained by taking into account as input signal a sinusoidal waveform superimposed to a nonzero offset. The nonzero offset is needed to polarize the diode with a forward bias. We examined the device response at different DC offset voltages as reported in Fig. 5.3.

The device shows a low-pass characteristics which can be explained by light level saturation effects due to the sinusoidal forcing signal and the offset. When the offset is increased, first the bandpass frequency and the input-output gain increase. Then, for further increases, saturation effects are dominant and the input-output gain decreases. Thus, increasing the offset effectively reduces the filtering effects until saturation effects become dominant.

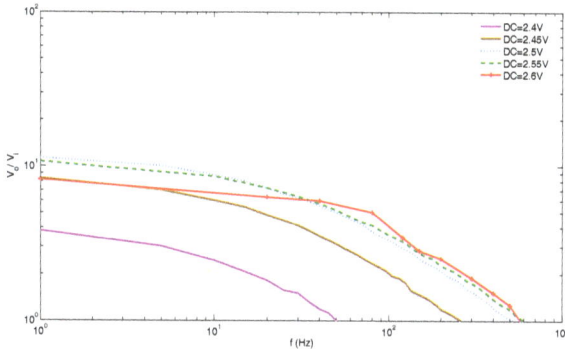

Fig. 5.3   Frequency response of the device shown in Fig. 5.2 for different values of the DC offset.

The nonlinear behavior of the device is also evident in Fig. 5.4 where the amplitude of the output signal is reported as a function of that of the input signal for different values of its frequency. As it can be noticed the input-output gain is linear only in a region of the characteristics and decreases as the frequency of the signal is increased.

5.2   The photo-controlled Chua's circuit

The nonlinear device described in the previous Section is used to introduce a light-controlled parameter in the Chua's circuit. To do this, we started from the SC-CNN implementation of the Chua's circuit discussed in Chapter 3 and we substituted a fixed-value resistor with the photoresistor as shown in Fig. 5.5. The photoresistor-controlled device of Section 5.1 is then used to control the circuit dynamics. In this way, on one hand, a further nonlinearity is included in the circuit and, on the other hand, a new parameter which can be controlled by an external decoupled signal is added.

The equations governing the photo-controlled Chua's circuit can be expressed in a dimensionless form by introducing in the Chua's equations (1.7) two implicit nonlinear functions of the light intensity L ($f(L)$ and $g(L)$) representing the effects of the introduced parameter on the state variables

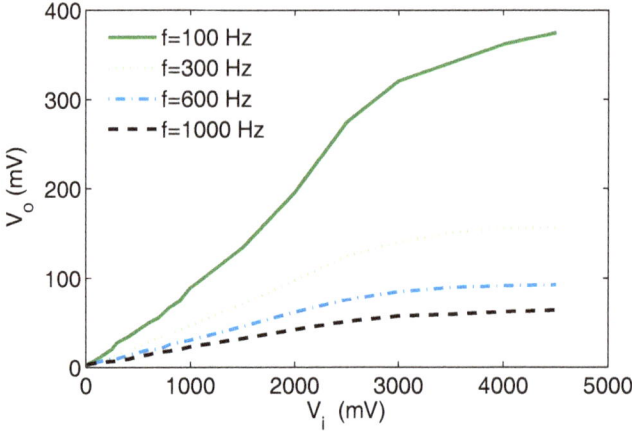

Fig. 5.4   Behavior of the device shown in Fig. 5.2 when the amplitude of the input signal is varied. Four different values of the frequency of the input signal have been considered.

of the circuit as follows:

$$\dot{x} = \alpha(y - h(x))$$
$$\dot{y} = f(L)x - y + g(L)z \qquad (5.1)$$
$$\dot{z} = -\beta y$$

The next Section will report experimental results showing as this further parameter is effective to control the behavior of the circuit.

5.3   Experimental results

In this section experimental results obtained with the light-control of the Chua's circuit are discussed. All the data have been acquired by using a data acquisition board (National Instruments AT-MIO 1620E) with a sampling frequency of $f_s = 200kHz$. The experiments refer to different ways of controlling the introduced circuit. When constant voltage signals are used, different attractors, periodicity, period-doubling bifurcations and route to chaos have been observed. When a periodic signal is used to control the photoresistor through the LED, other interesting phenomena such as switching between the right and left scroll at irregular time intervals have been observed. This switching behavior can be regularized by feedback as experimental results demonstrated.

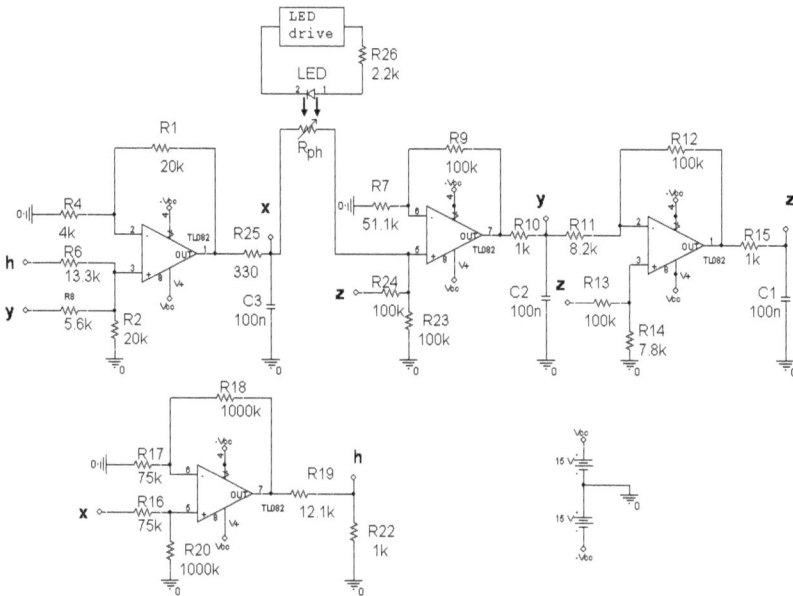

Fig. 5.5 Schematics of the Chua's circuit with the photoresistor-diode device. The following parameters are assumed: $\alpha = 9$, $\beta = 14.286$, $m_0 = -0.14$, $m_1 = 0.28$.

As a first experiment, a constant input to the photoresistor-LED device is taken into account. As the voltage is varied, the light intensity also changes. This in turn changes the value of the resistance associated to the photoresistor. Experimental results obtained using such input lead to the conclusion that varying this parameter the behavior of the circuit can be modulated and different attractors can be obtained. Moreover, for a suitable value of the input parameter, a sequence of period-doubling bifurcations leading to chaos are observed. Figure 5.6 shows the bifurcation sequence obtained when the light intensity is increased by varying the DC input from 2.433 V to 2.448 V. First, a periodic behavior is obtained, then a period-doubling sequence is observed, then the circuit exhibits a single scroll and finally a double scroll attractor. As the light intensity is further increased, the double scroll collapses to an external stable limit cycle which is the only solution observed for high values of the constant DC input.

In a second set of experiments, a sinusoidal voltage signal is feed as input to the LED. In order to provide polarization to the diode an offset is added to this signal to form the real input of the circuit: $v_i(t) = V_{DC} +$

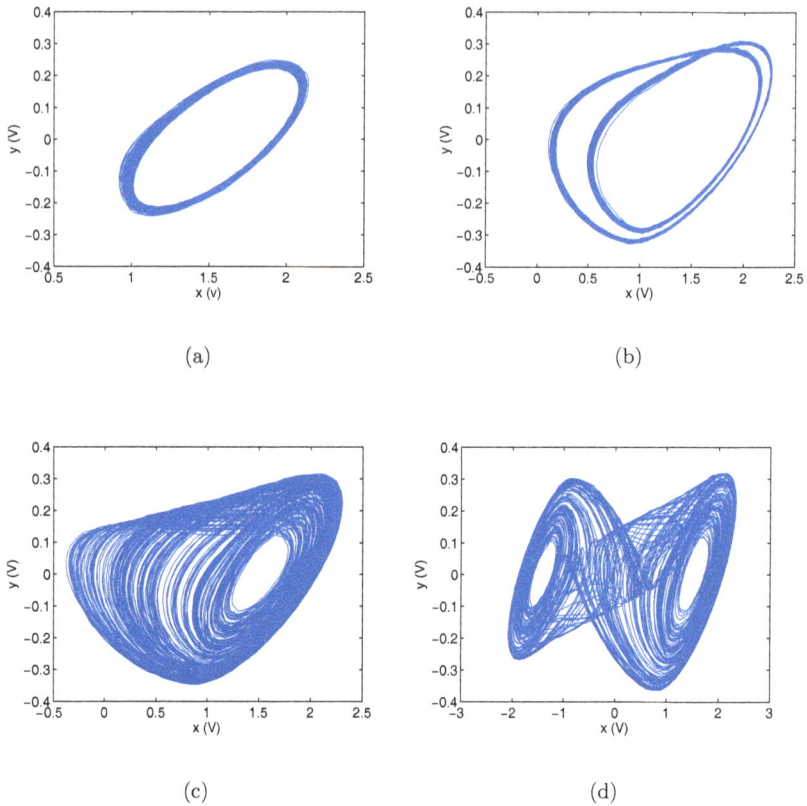

Fig. 5.6   Attractors obtained for different values of the DC constant input. (a) $DC = 2.433V$; (b) $DC = 2.443V$; (c) $DC = 2.447V$, (d) $DC = 2.448V$.

$A\sin(2\pi ft)$. The behavior of the Chua's circuit depends on a complex interplay between all the parameters which characterize the input signal (amplitude, frequency and offset). For a suitable range of parameters the circuit shows chaotic behavior. We restrict here our discussion to some interesting chaotic behaviors shown by the circuit. For given values of the amplitude and offset of the signal ($V_{DC} = 1.2V$ and $A = 80mV$), varying the frequency from $f = 15Hz$ to $f = 35Hz$ the circuit behavior changes: double scrolls are observed for frequency values $f < 25Hz$; single scrolls are obtained for frequency values $f \geq 29Hz$; for intermediate values switching between the left and right single scroll is observed. In Fig. 5.7 different attractors obtained at different frequencies of the input signal are shown. To better visualize the switching behavior, Fig. 5.8 reports the waveforms of the x variable, showing how the circuit behavior switches in an irregular way from the right scroll to the left one and viceversa. Therefore, we can conclude that, even if the filtering effects of the photoresistor-diode are reduced with higher values of the DC offset, a low frequency signal produces intermittency in the chaotic circuit which operates at frequencies up to some kilohertz, *i.e.* in the classical Chua's circuit working range.

The intermittency has been revealed in a wide set of experimental observations. In particular, the time interval when intermittency occurs is highly irregular and has been conjectured for it a chaotic behavior.

We also investigated the effects of other periodic signals, such as triangular or square waves. Compared to sinusoidal signals, we found the same effects but at lower frequencies. This means that the harmonic components which are present in the square or triangular waves are playing an important role by compensating the effect of reducing the main harmonic.

Starting from the switched scroll attractors, we investigate the question whether it is possible to stabilize one of the observed scrolls. We found that this is indeed possible and can be achieved by using feedback.

Figure 5.9 describes how feedback is realized in the circuit. The feedback signal is built from the y state variable of the circuit by summing it to the driving forcing signal. This feedback signal is injected in the circuit by using the LED driving the photoresistor. The control feedback parameter is the feedback resistor R_f.

Figure 5.10 shows the effect of feedback on the circuit. The circuit is driven by a sinusoidal signal such that without feedback the switching behavior shown in Fig. 5.8 is observed. From the comparison of Fig. 5.10 and Fig. 5.8 it is evident that the feedback has a stabilizing effect: without feedback the circuit behavior switches from scroll to scroll, while feedback stabilizes the behavior into a double scroll attractor. Thus, with feedback control the intermittent behavior is regularized and the attractor is exactly a classical double scroll attractor.

We further investigate this stabilizing effect by increasing the feedback strength (*i.e.*, slowly decreasing R_f). This leads to stabilizing a single scroll attractor and a further increase lead to a stable limit cycle. Figure 5.11 reports three different attractors obtained by increasing the feedback strength. In particular, Fig. 5.11(a) refers to the same case shown in Fig. 5.10.

In this Chapter, we have described the dynamics of a Chua's circuit

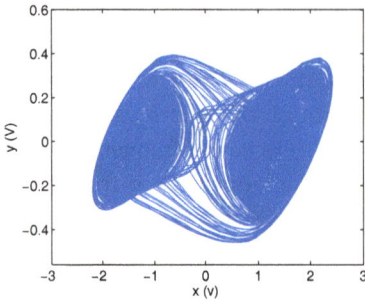

(a)

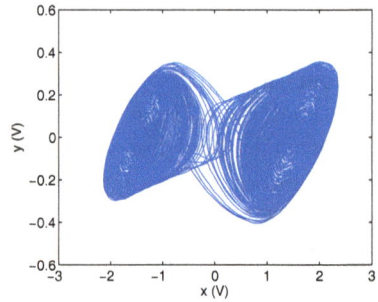

(b)

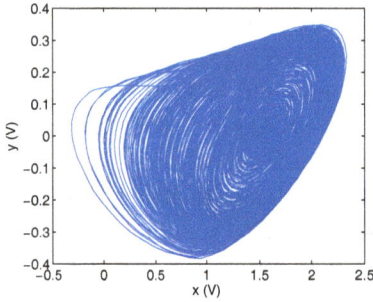

(c)

Fig. 5.7 Attractors obtained for different values of the frequency of a sinusoidal input signal of the form: $v_i(t) = V_{DC} + A\sin(2\pi ft)$. (a) $f = 15Hz$; (b) $f = 25Hz$; (c) $f = 29Hz$.

in which the introduction of a photoresistor coupled with a LED allows to control the behavior of the circuit by the light emitted by the diode. The main conclusion that can be derived from the experiments performed on the circuit driven by different types of control signal is that different attractors, period-doubling bifurcations and route to chaos can be obtained. In particular, when periodic signals are used, interesting phenomena such as switching between the right and left scroll at irregular time intervals arise. This switching behavior is regularized by feedback as experimental results demonstrated.

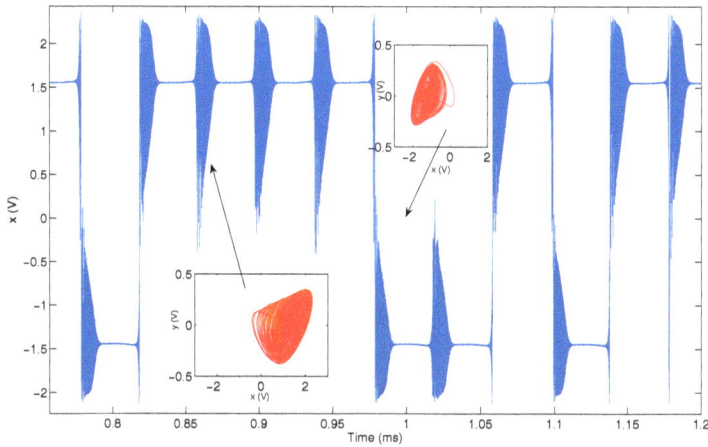

Fig. 5.8 Behavior of the system when driven by a sinusoidal input signal with $f = 25Hz$, $V_{DC} = 1.2V$ and $A = 80mV$. An unstable behavior switching from right to left (and viceversa) scroll attractor is observed.

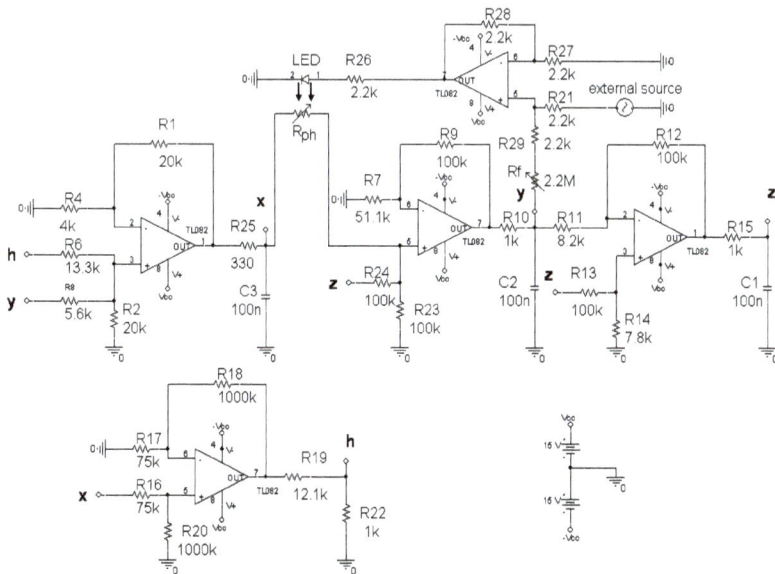

Fig. 5.9 Scheme of the circuit with feedback.

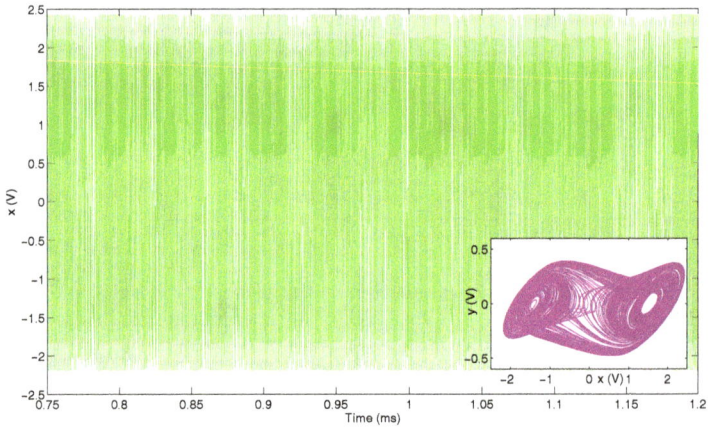

Fig. 5.10   Stabilizing effect of the feedback: trend of the x state variable.

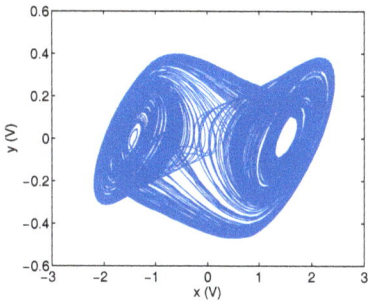

(a)

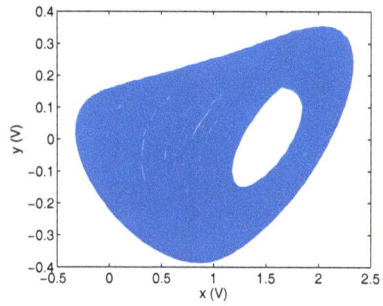

(b)

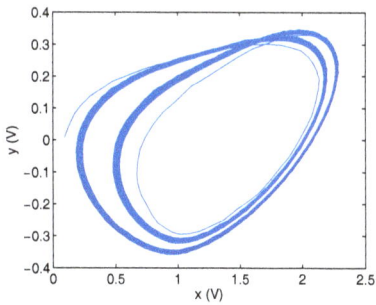

(c)

Fig. 5.11   Stabilizing effect of the the the feedback: different stable attractors obtained by increasing the feedback strength (*i.e.*, slowly decreasing the feedback resistance R_f).

Chapter 6

Programmable Chua's circuit

In this chapter we discuss an implementation of the Chua's circuit based on a programmable analog electronic device. The design proposed is based on the CNN-implementation described in Chapter 3.

Analog computation is a fascinating paradigm in complexity studies and bio-inspired electronic applications. However, the main drawback of analog computers is the difficulty of programming them. Cellular Neural Networks (CNNs) are the most famous example of analog processors used in a range of applications from image processing to complex system emulation. A new paradigm for the realization of programmable analog processors was recently introduced starting from the digital analog FPGA (Field Programmable Gate Array). Although this new class of devices called FPAA (Field Programmable Analog Array), [D'Mello and Gulak (1998)], is in the early stages of development and therefore of applications, it seems to have a promising future as a paradigm for rapid and reprogrammable analog design. The first devices include only a small number of programmable blocks, but have found several applications in the field of signal processing (filtering) [Birk (1998)].

The experimental results shown in this Chapter, obtained using Anadigm FPAA devices [Anadigm (2003)], show the suitability of the FPAA-based approach for chaotic circuit implementation and highlight several advantages of the approach: the design and implementation phases are very simple, and the circuit is totally programmable; in fact FPAA offers the possibility to change all the circuit parameters on the fly. For these reasons the FPAA implementation of Chua's circuit is in itself an interesting new realization: a fully programmable Chua's circuit is obtained.

Digital circuits are usually used when high-accuracy, high-complexity signal processing algorithms have to be implemented. This approach is typically applied to low-frequency signals when power consumption is not critical. When, on the contrary, high-frequency signals are involved and low power dissipation is required, the analog approach is more appealing. The drawback of the analog approach is related to the accuracy of the circuits and their reprogrammability. The new FPAA technology, as for its digital counterpart FPGA, provides a basic instrument for the development of dynamically reconfigurable analog circuits. Using this kind of device it is possible to reprogram the entire circuit dynamics, keeping the structure fixed but changing the parameters. The reprogrammability features of FPAA can also be used to adapt the circuit to changing external conditions due to noise or changes in the operating conditions of the system being controlled. The circuit configurations can be changed at a low level, where components such as operational amplifiers, capacitors, resistors, transconductors, and current mirrors can easily be fixed and connected, and also at a high level. In the latter case, user-friendly tools, for example to design audio amplifiers, are available in order to reduce the time to market for products. In a way, FPAA is a new version of an analog computer, maintaining the same target of providing analog circuits with flexibility. Therefore, the two main characteristics of FPAA are the possibility to translate complex analog circuits into a set of low-level functions and the capability to place analog circuits under real-time software control within the system. For these reasons FPAA is used here to implement chaotic circuits with programmable features.

6.1   The Development System

The device used in the folllowing is the AN221E04 FPAA produced by Anadigm [Anadigm (2003)], integrated on the AN221K04 development board produced by the same company. The software development tool is AnadigmDesigner2.

The core of AN221E04 is a two-by-two matrix of blocks named CAB (Configurable Analog Blocks) that can be connected with each other and with external I/O blocks. Fig. 6.1 gives a scheme of the Anadigm device.

Each CAB contains a digital comparator, an analog comparator, two operational amplifiers and a series of capacitors. FPAA technology is, in fact, mainly based on switched capacitor technology. The CAB blocks are surrounded by the other elements of the device. One section is dedicated to clock management, another to signal I/O and a digital section is devoted

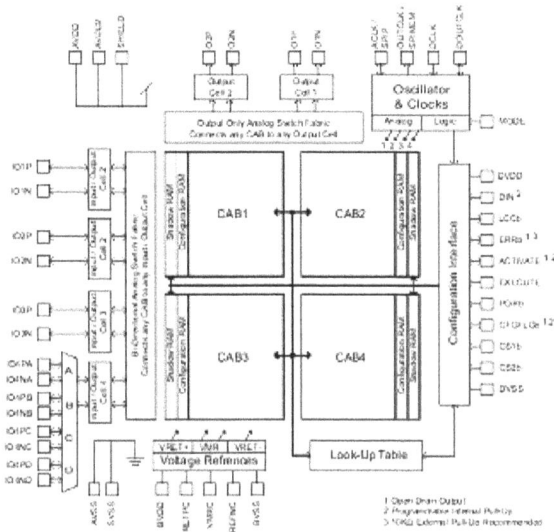

Fig. 6.1 Block scheme of the AN221E04 device.

to IC configuration and dynamic reprogrammability. The digital section is based on a look-up table mapping the interconnections inside the IC. The look-up-table allows the FPAA to be dynamically reprogrammed. It is, in fact, possible to connect the AN221E0 with an external microcontroller and to change the values of the look-up-table on the fly. These values will be applied at the next clock cycle. Other features characterizing the AN221E0 are not described here because they are needed in the experiments.

6.2 Anadigm Software Development Tool

The software development tool allows the FPAA to be connected with the I/O ports and the desired circuit to be designed using pre-defined blocks. There are a large number of blocks called CAMs (Configurable Analog Modules) with different functions and therefore resource requirements. Here only those blocks that will be used in our experiments are described.

The first, and the simplest, is the GAININV block that allows us to fix a gain, (in the range 0.01÷100), and an internal clock for switch driving.

A further block is SUMINV. This block integrates its inputs by using a fixed time constant and outputs the resulting signal. Here again it is necessary to fix the clock time for the internal switching. The transfer function implemented by the block is the following:

$$V_{Out}(s) = \frac{\pm K_1 V_{Input1}(s) \pm K_2 V_{Input2}(s) \pm K_3 V_{Input3}(s)}{s}$$

where the three constants are given by:

$$K_i = \frac{f_c C_i}{C_4}$$

Another CAM is SUMFILTER, whose transfer function is:

$$V_{Out}(s) = \frac{2\pi f_0 [G_1 V_{Input1}(s) + G_2 V_{Input2}(s)]}{s + 2\pi f_0}$$

where:

$$f_0 = \frac{f_c}{\pi} \cdot \frac{C_5}{2C_4 + C_5}$$

$$G_i = \frac{C_i}{C_5}$$

This block, similar to the previous one, apart for the presence of a non-zero pole, allows the basic cell of a CNN to be implemented reducing the number of inputs as compared with SUMINTEGRATOR. In fact, this block implements an RC integrator (see Chapter 2) and can be used to implement the core of a CNN cell.

The VOLTAGE block provides a voltage reference and is used to realize the bias term.

6.3   Design of the Chua's circuit

As it has been already shown, it is possible to implement Chua's circuit by using CNNs (Chapter 3). However, since the internal voltage signals of the FPAA are bounded, the dynamic range of the state variables should be examined to guarantee the suitability of the FPAA-based implementations.

In fact the state space variables x, y, z, of the circuit vary in the ranges (-2.2 ; 2.2) , (-0.5 ; 0.5) and (-3 ; 3) respectively. Due to the fact that the FPAA-based circuit saturates if a voltage greater than 2 V is reached, the Chua's equations given in their original form cannot be used. In order to overcome this problem, new state variables are adopted: the variables x and z are rescaled by a scale factor k as follows:

$$\begin{cases} X = \frac{x}{k} \\ Y = y \\ Z = \frac{z}{k} \end{cases} \tag{6.1}$$

The equations of the Chua's circuit therefore takes the following form:

$$\begin{cases} \dot{X} = \frac{\alpha}{k}[Y - h(kX)] \\ \dot{Y} = kX - Y + kZ \\ \dot{Z} = \frac{-\beta}{k}Y \end{cases} \tag{6.2}$$

6.4 Circuit implementation and experimental results

Figure 6.2 shows the blocks used inside the AnadigmDesigner2 environment implementing the Chua's circuit [Caponetto *et al.* (2005)]. Each block is characterized by a number, from 1 to 7 and the function of each single block is described below.

Blocks (1)-(3), are SUMFILTER blocks while blocks (4)-(7) are GAININV blocks.

The output of the SUMFILTER blocks is the three state space variables X, Y, and Z. Some of the circuit gains were implemented directly inside the SUMFILTER blocks, while others were fixed using the GAININV blocks. This choice was due to two considerations. The first is related to the limited range of gains settings in the SUMFILTER block. This range is too small for the gains α and β of the Chua's circuit. The second consideration is related to the accuracy of the gains that can be set in the SUMFILTER blocks. This accuracy is less than that achieved by using the GAININV block.

It must be stressed that all the circuit gains could be implemented by a single GAININV block, but with this approach the number of operational amplifiers increases and does not allow the Chua's circuit to be implemented using only one device.

GAININV block (6) plays an important role: by varying this gain, denoted as g, it is possible to change the dynamic behavior of the circuit following a period-doubling route to chaos.

GAININV blocks (4) and (5) are used to implement the output nonlinearities of equation (6.2). In this block the 2V saturation limit is used to implement the nonlinearity. The Chua's circuit nonlinearity saturates the

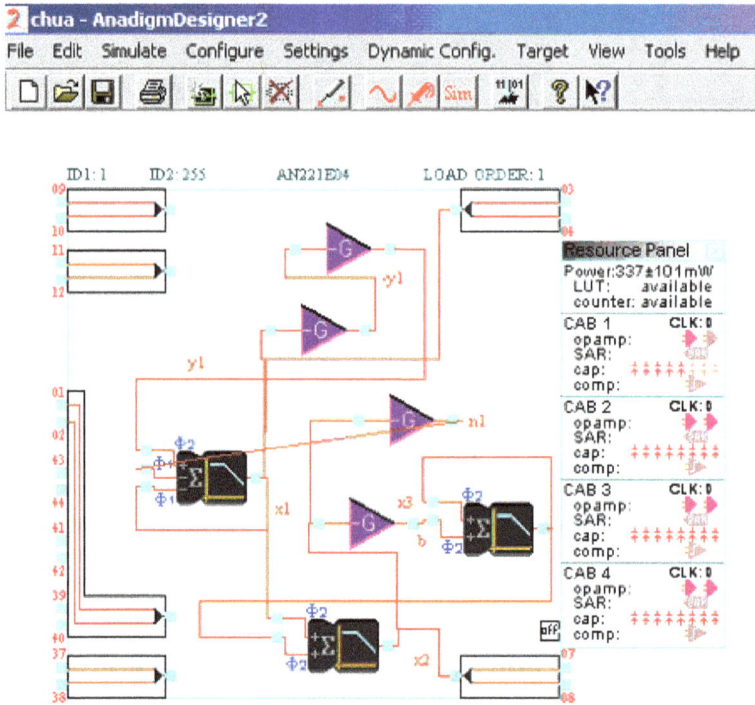

Fig. 6.2 Scheme of Chua's circuit using the AnadigmDesigner2 environment.

signal at 1V. Therefore, in order to use a GAININV block, the signal was pre-amplified by a factor of 2 and then rescaled in the range [0 1V].

The values of the circuit parameters were chosen to implement the well known Chua's circuit attractors. However, these values slightly differ from the theoretical ones. This is due to the fact that a tolerance error close to 20% may appear in some blocks. The parameters fixed for the CAM implementing the block just described are given in Table I.

The circuit was downloaded onto the AN221K04 development board shown in Fig. 6.3.

Table 6.1   CAM parameters implementing the scheme shown in Fig. 6.2.

CAM Type	CAM number	Tuned parameter	Parameter Value
	1	Corner frequency	0.4
		Gain1	2
		Gain 2	1
		Gain 3	1.5
SUMFILTER	2	Corner Frequency	0.4
		Gain1	2
		Gain 2	2
	3	Corner Frequency	0.4
		Gain 1	1
		Gain 2	1
	4	Gain	4
GAININV	5	Gain	0.5
	6	Gain	g
	7	Gain	6

The double scroll was obtained for $g = 3.18$. Fig. 6.4(a) shows the double scroll attractor in the $x - y$ plane, while Fig. 6.4(b) shows the state variables x and y.

Further experiments were devoted to investigate the different dynamics of the Chua's circuit. The bifurcation parameter g was varied in the range 2.6-3.18, showing a large spectrum of behaviors ranging from stable equilibrium points to chaotic attractors. The well known period doubling route to chaos was experimentally observed. In Fig. 6.5 several attractors are shown.

By fixing $g = 2.6$ an equilibrium point is obtained, as shown in Fig. 6.5(a). With $g = 2.63$ and $g = 2.7$ a limit cycle appears, (Fig. 6.5(b)-(c)). By further increasing the parameter g period doubling bifurcations of the limit cycle are observed until the single scroll chaotic attractor develops (Fig. 6.5(d)-(f)). A higher value for parameter g leads to the double scroll shown in Fig. 6.4. It should be noticed that as the Chua's circuit is sym-

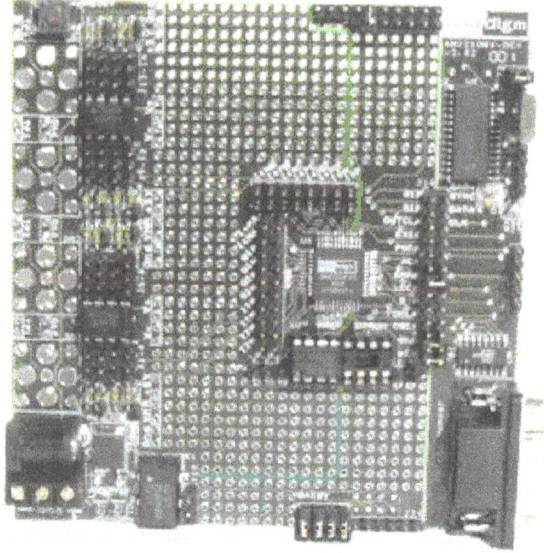

Fig. 6.3   The AN221K04 evaluation board.

metric, in Fig. 6.5(a)-(f) there also exists the corresponding symmetrical attractor.

The experimental setup based on the Anadigm FPAA device provides the possibility of exploring the set of different attractors in a very simple way. It is, in fact, possible to create a graphic user interface with a slide controlling the bifurcation parameter of the Chua's circuit, as shown in Fig. 6.6. This feature is very appealing for two reasons: first, it allows us to build a low-cost introductory kit on the basic properties of chaotic circuits for educational purposes; then, it permits us to explore the entire range of parameters in a very simple way.

To conclude this Chapter, further experimental results referring to the synchronization of two Chua's circuits are given. The scheme in Fig. 6.2 was reduced in order to allow the implementation of two different Chua's circuits onto the same FPAA device. The well-known system decomposition scheme was adopted. The system decomposition we adopted is shown in Fig. 6.7. Experimental results showing the synchronization between the two circuits are given in Fig. 6.8.

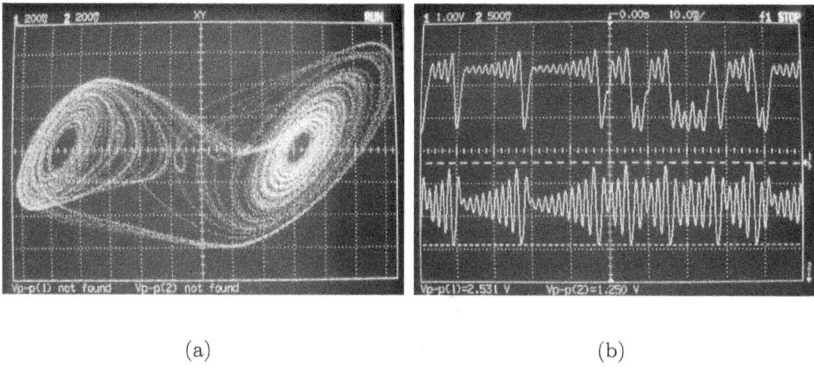

(a)   (b)

Fig. 6.4   (a) Projection of the double scroll attractor in the $x - y$ plane ($g = 3.18$). (b) Trend of the state variables x (up) and y (down) ($g = 3.18$).

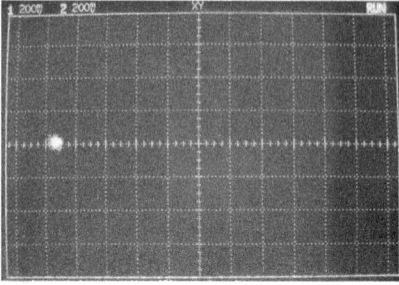

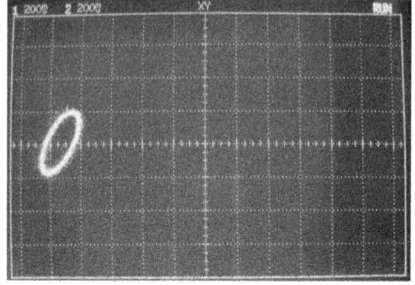

(a)   (b)

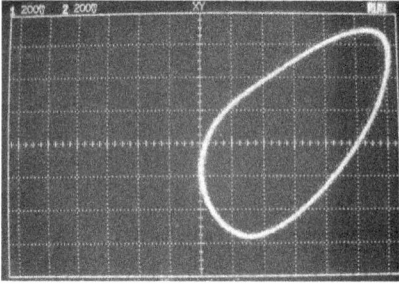

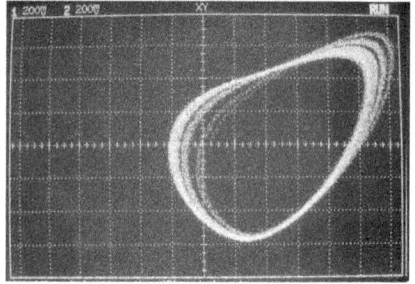

(c)   (d)

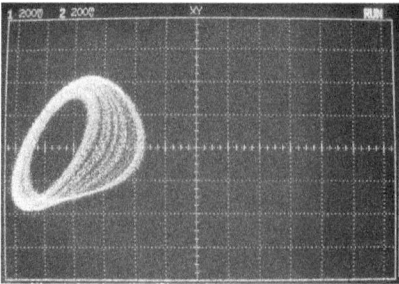

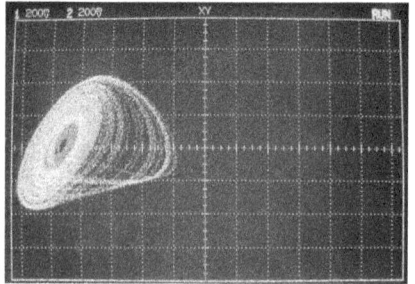

(e)   (f)

Fig. 6.5   Set of experimental attractors obtained with different values for the parameters g: (a) $g = 2.6$; (b) $g = 2.63$; (c) $g = 2.7$; (d) $g = 2.8$; (e) $g = 2.9$; (f) $g = 3$.

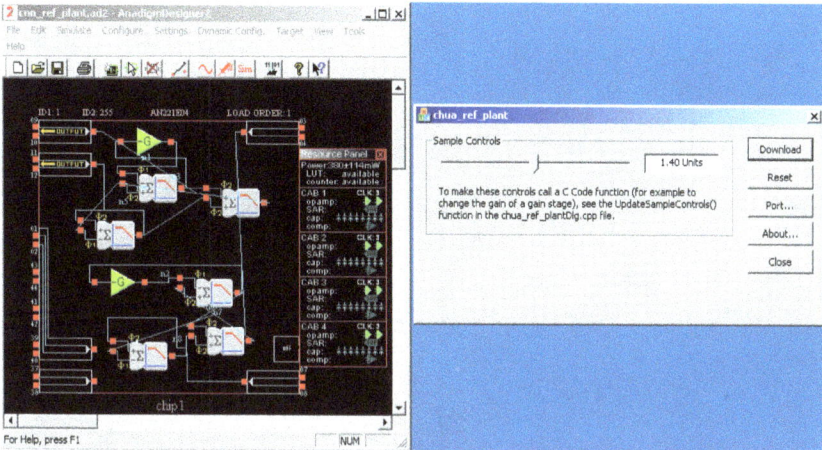

Fig. 6.6   Graphic user interface to explore bifurcations in the FPAA Chua's circuit.

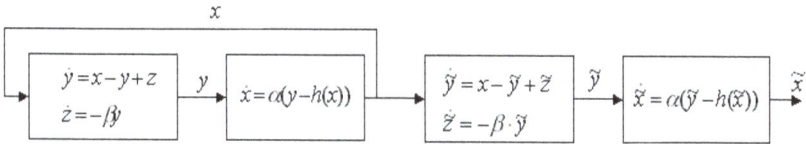

Fig. 6.7   Scheme of the system decomposition adopted to synchronize two FPAA Chua's circuits.

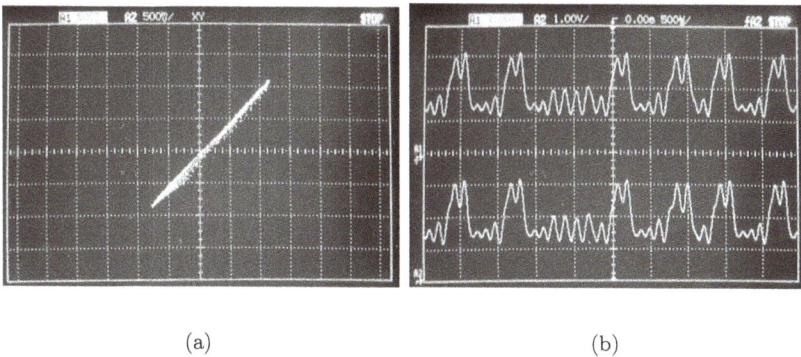

(a)   (b)

Fig. 6.8   Experimental synchronization of two FPAA Chua's circuits: synchronization curve and comparison between variable x of master and slave circuits.

Chapter 7

Switched capacitor-based Chua's circuit

In this Chapter the design of an integrated Chua's circuit [Arena *et al.* (2004)] is presented. Also in this case the starting point for the design is the CNN-based implementation described in Chapter 3 . As already said, several investigations on the global behavior shown by systems made of many coupled Chua's circuit have been realized. Studies on propagation of impulsive information in chains of Chua's circuits [Perez-Munuzuri *et al.* (1992)], global chaotic behavior [Dabrowski *et al.* (1993)], and on the effects of dissymmetries in the connection parameter on the global behavior of the system [Arena *et al.* (2000)] are only a few examples of the topics addressed. In particular in the last example, the number of Chua's cells needed is very high. Such investigations can only be experimentally supported by an integrated silicon chip which contains a large number of Chua's cells.

The design proposed in this Chapter is based on switched-capacitor (SC) techniques [Gregorian (1986)], and is oriented to a small area consumption. The aim is to implement a cell which can constitute the basic element to build a chip implementing an array of Chua's cells and exploit the connectivity of the CNN implementation to perform experimental investigations on arrays of Chua's circuits.

The implementation presented aims at reproducing the dynamics generated by the Chua's equations by assuming 1V as dimensional unit for the x, y, z variables with following time rescaling:

$$t' = t \cdot \tau.$$

Concerning past implementations, the IC Chua's circuit presented in [Cruz and Chua (1993)] is based on the original implementation of the Chua's circuit with the inductor realized with a capacitor and a gyrator, while the circuit reported in [Delgado-Restituto and Rodriguez-Vazquez

(1993)] is based on Voltage Controlled Current Sources. Using State Controlled CNN allows a flexible design and the possibility of changing the implemented dynamics with a suitable choice of the template coefficients.

Figure 7.1 shows the active RC circuit, already shown in Chapter 2, rearranged in a nested structure. The active RC network makes use of a small number of discrete passive components and is the starting point in the SC circuit design procedure.

7.1   Design of the switched capacitor Chua's circuit

Mapping the dynamics of the Chua's equation with the equations describing the circuit illustrated in Fig. 7.1, the following relations between component values and Chua's circuit parameters can be easily derived:

$$\tau = C_2 R_6;$$
$$\alpha = \frac{C_2 R_6}{C_1 R_5};$$
$$\beta = \frac{C_2 R_6}{C_3 R_9};$$
$$m_1 = \frac{R_5}{R_4};$$
$$m_0 = m_1 - \frac{R_1 R_5}{R_2 R_3}.$$

The corresponding SC implementation is obtained by substituting each resistor in the active RC circuit, with an equivalent switched capacitor block [Gregorian (1986)]. In the SC technique, resistors are implemented by means of capacitor and switches, connected in a way that depends on the adopted mapping between continuous and discrete time domain. Moreover, the design of the operational amplifiers for SC circuits is simplified because they do not need a high power output stage. In fact they have only capacitive loads, so high output currents are not required. Therefore Operational Transconductance Amplifiers (OTAs) can be used for on-chip signal processing. This choice reduces chip area consumption and improves the bandwidth.

Figure 7.2 shows the SC implementation, where the expression of the parameters of the system, in terms of the component values, is given by:

$$\alpha = \frac{1}{T}\frac{C_6}{C_5}\tau;$$
$$\beta = \frac{1}{T}\frac{C_{10}}{C_{12}}\tau;$$
$$m_1 = \frac{C_4}{C_6};$$
$$m_0 = m_1 - \frac{C_2 C_3}{C_1 C_6};$$
$$\tau = T\frac{C_9}{C_{11}} = T\frac{C_9}{C_7} = T\frac{C_9}{C_8}$$

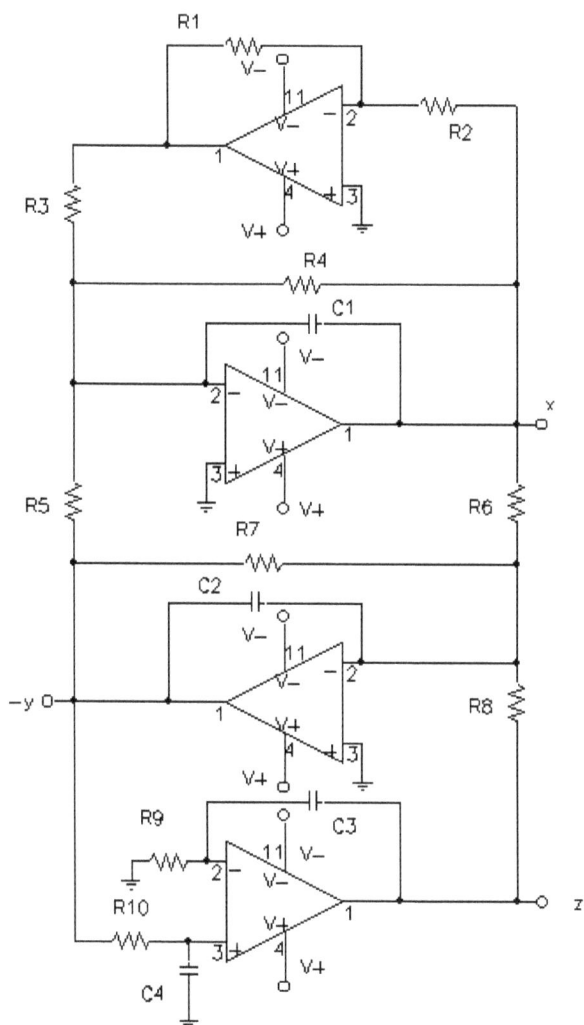

Fig. 7.1 The RC active implementation of the Chua's circuit based on State Controlled CNNs.

where T is the clock period.

The time rescaling factor can be varied in a simple way by changing the clock frequency, since the choice of capacitors C_9 and C_{11} fix only the ratio between T and τ. This is a key point of the SC implementation: the

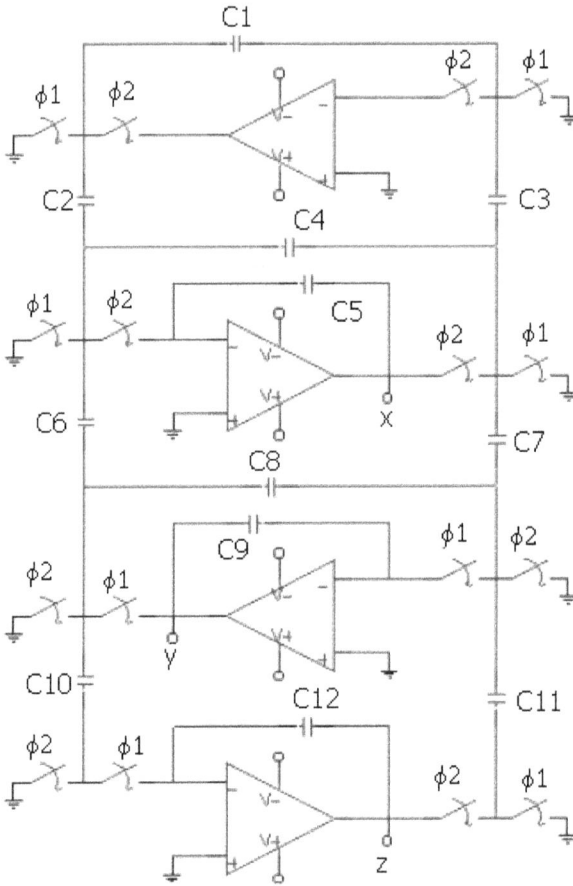

Fig. 7.2　The SC implementation of the Chua's circuit.

frequency of the chaotic oscillations can be varied over a wide range.

　　To obtain the possibility of generating several attractors of the Chua's system by changing parameters α and β two sets of capacitors connected in parallel to a fixed one (C_5 for the α parameter and C_{12} for the β parameter in Fig. 7.2 are included in the system. The capacitor to be actually connected is selected by an external input. This choice provides the possibility of changing in a discrete way the parameters of the circuit, but it avoids the need of external, variable components as in [Cruz and Chua (1993)], or external current references as in [Delgado-Restituto and Rodriguez-Vazquez

(1993)].

7.2   Simulation results and layout of the circuit

In this Section simulation results obtained by choosing one of the attractors that can be generated by the Chua's circuit are reported. Moreover, the layout of the cell is presented.

The values of the components of the SC circuit shown in Fig. 7.2 have been selected in order to minimize the area consumption. To address the minimum area requirement, the minimum capacitance value allowed by the technology has been chosen for the capacitors C_1, C_2, C_7, C_8, C_{11} and C_{12}.

The design phase has been accomplished by taking into account the parameters for the double scroll strange attractor (Section 1.5.1) and fixing the clock frequency at $\frac{1}{T} = 1MHz$ and the constant time normalization factor at $\tau = 10^{-5}s$, even if as discussed above other attractors can be obtained and the time normalization factor can be changed. In particular the following parameters have been adopted:

$$\alpha = 9; \beta = 12.82; m_0 = -\frac{1}{7}; m_1 = \frac{2}{7}.$$

This allows us to select the remaining components. The results of the simulation of the circuit are shown in Figs. 7.3 and 7.4, with the following choice of component values: $C_1 = 0.1pF$; $C_2 = 0.1pF$; $C_3 = 0.3pF$; $C_4 = 0.2pF$; $C_5 = 0.78pF$; $C_6 = 0.7pF$; $C_7 = 0.1pF$; $C_8 = 0.1pF$; $C_9 = 1pF$; $C_{10} = 0.128pF$; $C_{11} = 0.1pF$; $C_{12} = 0.1pF$;

In particular, Fig. 7.3 shows the trends of the y and x variables, while Fig. 7.4 illustrates the *double scroll strange attractor* in the $x - y$ phase plane.

A further simulation result, shown in Fig. 7.5, confirms the suitability of the approach for different clock frequencies. The clock frequency can be adjusted to achieve chaotic behavior at different frequencies. Figure 7.5 shows the trend of the variable x of the Chua's circuit, generated by setting four different clock frequencies. Satisfactory results have been achieved in the range 500Hz-2MHz.

The layout of the Chua's cell, consisting of three blocks, is illustrated in Fig. 7.6. The upper block consists of four OTAs, the middle block consists of switched capacitors and the lower block includes the phase generator and the current reference. The technology adopted is CMOS AMS 0.8 μm.

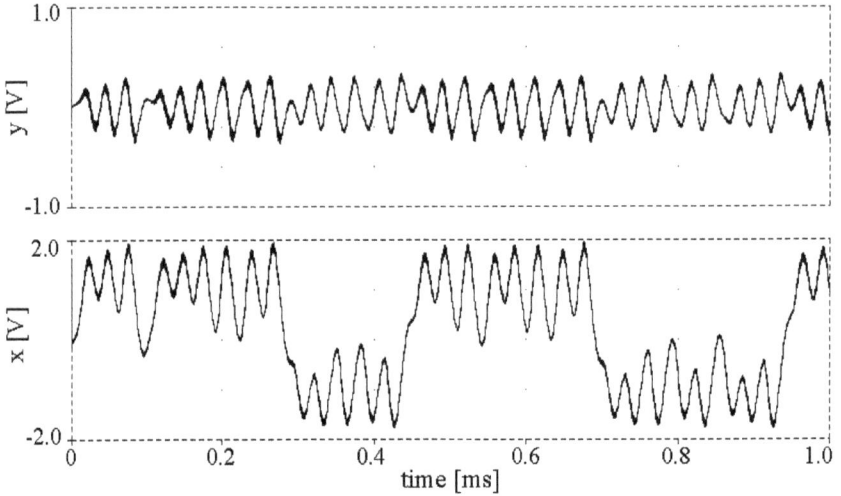

Fig. 7.3   Results of the simulation of the SC circuit: (a) variable y; (b) variable x.

With this implementation the power dissipation is $3mW$.

7.3   Experimental results

The prototype of the IC Chua's circuit described in this Chapter has been realized in a 48 pin chip containing other prototypes under study in our laboratories. A photograph of this chip is shown in Fig. 7.7. The Chua's circuit occupies a small portion ($200\mu m$ x $200\mu m$) of the whole silicon area and requires 12 pins: three pins are devoted to the dual voltage supply (± 2.5V) and ground reference, three pins are the state variables of the Chua's circuit, one pin is the clock signal, and five pins are the control signals needed to change the α and β parameters.

The chip operates in the double scroll region by setting the capacitance values as in simulation.

Figures 7.8 and 7.9 show the projection of the double scroll and the single scroll attractor onto the $x - y$ phase plane, when the clock frequency is fixed to the value $f_c=100kHz$.

It has been experimentally verified that by varying the clock frequency, chaotic oscillations at different frequencies can be obtained. In the real

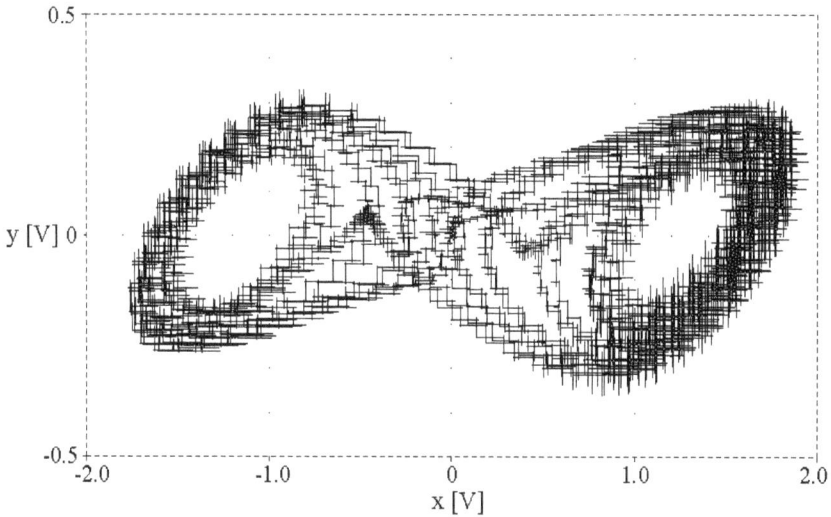

Fig. 7.4   Projection of the Double Scroll Chua attractor onto the $x - y$ phase plane.

implementation, the suitable range for the clock frequency is 500Hz-500kHz. The agreement between simulated and experimental results is in the order of 10%.

7.4   A cell to experimentally characterize discretely-coupled Chua's circuits

The Chua's circuit is an universal paradigm for studying nonlinear dynamics and chaos in discretely-coupled arrays. As a function of its parameters, distinct single cell dynamics can be obtained [Munuzuri *et al.* (17–50)], leading to distinct behaviors of the system consisting of discretely-coupled Chua's circuits. For instance, it can be shown that for different parameter values the Chua's circuit admits:

- Three equilibrium points, two of them stable and one unstable. In this case, when the first few circuits of a one-dimensional array of resistively coupled circuits switch from a stable equilibrium point to the other stable one, the behavior of the array is characterized by a travelling wave solution.
- One stable equilibrium point, in this case the system is *excitable*, since

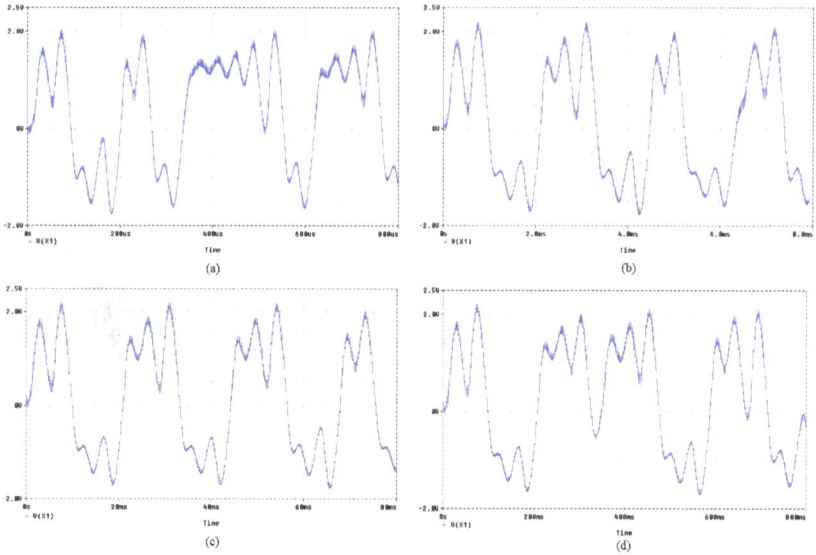

Fig. 7.5   Waveforms of variable x (Double Scroll Chua attractor) for different clock frequencies: (a) f_c=500kHz; (b) f_c=50kHz; (c) f_c=5kHz; (d) f_c=500Hz.

a sufficiently large perturbation on the initial state leads to a large excursion in the phase plane before settling down to the equilibrium state.

• One unstable equilibrium point, in this case the system is an *oscillating medium*, characterized by a relaxation limit cycle.

These behaviors can be obtained with the circuit presented in this chapter by introducing a slight modification of the nonlinearity. In fact, in order to obtain all of these dynamics, an asymmetrical nonlinearity has to be considered.

When the single cell dynamics is characterized by a bistable solution, a traveling wave is evident in arrays of Chua's circuits. In this case there is a transition between the two stable states of the circuit with a constant velocity. In steady state all the circuits evolve towards the same stable point and the velocity of the traveling wave is function of the diffusion coefficient. Moreover, for a critical value of the diffusion coefficient the wave fails to propagate in the medium. Numerical examples of traveling waves propagating in arrays of Chua's circuits are reported in [Munuzuri *et al.* (17–50)].

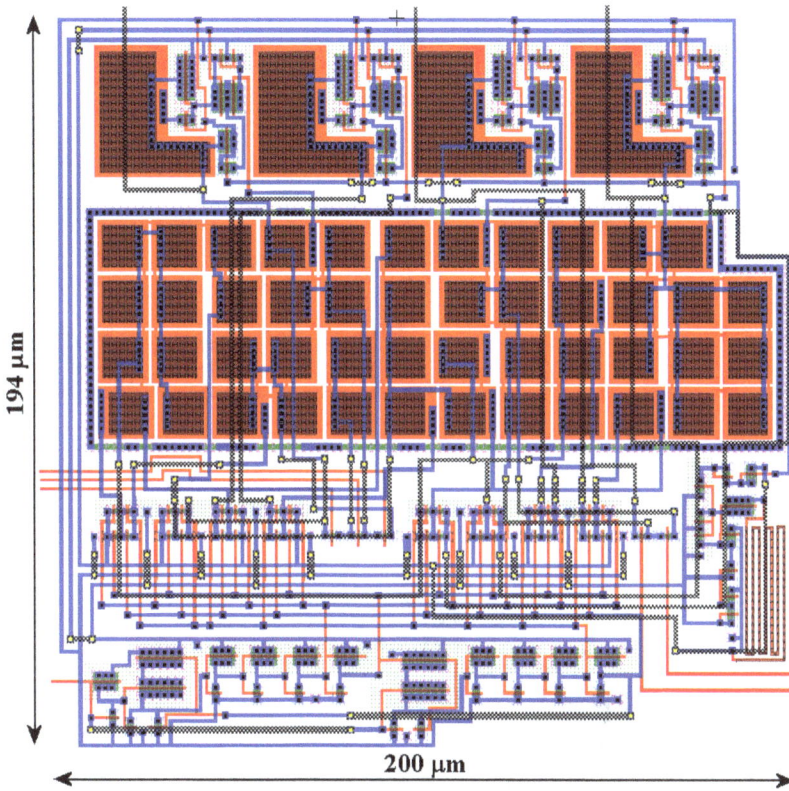

Fig. 7.6   Layout of the SC Chua's cell.

On the other hand, when the single cell dynamics is that of an excitable system or an oscillating medium, spiral waves, target waves and auto waves can be observed in two-dimensional arrays of Chua's circuits. Wave patterns are evident also in three-dimensional arrays [Munuzuri *et al.* (17–50)]. The birth of spiral waves in such systems is particularly important and can be related to similar phenomena occurring in cardiac tissues. In fact, spiral waves have the highest possible frequency for waves spontaneously propagating in the medium. This means that, once a spiral wave begins to propagate, it annihilates all other structures present in the medium. In the cardiac tissue the normal functioning dynamics is represented by a concentric wave propagating from the atrial sinus: spiral waves are therefore abnormal and represent potentially dangerous states of cardiac activity.

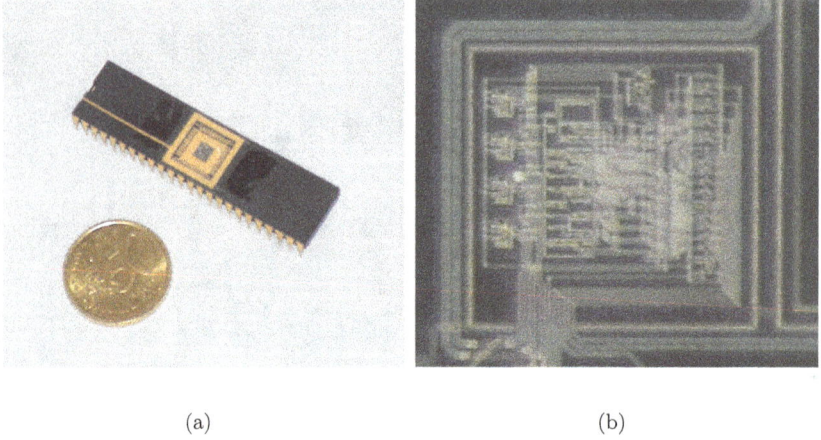

(a)   (b)

Fig. 7.7   Photograph of the chip.

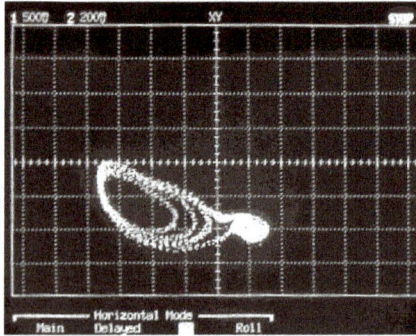

Fig. 7.8   Experimental results: projection of the double scroll strange attractor onto the $x - z$ phase plane.

A further result obtained with arrays of Chua's circuits refers to pattern formation and is illustrated in [Munuzuri *et al.* (17–50)]. In this case a reduced-order Chua's circuits is considered. Other studies on arrays of Chua's circuits are reported in [Perez-Munuzuri *et al.* (1992); Dabrowski *et al.* (1993)].

In all of these examples the cells are assumed to be identical. Even more interesting behaviors arise in arrays of non-identical Chua's circuits as illustrated in [Arena and Fortuna (2000)]. In this case an array of chaotic

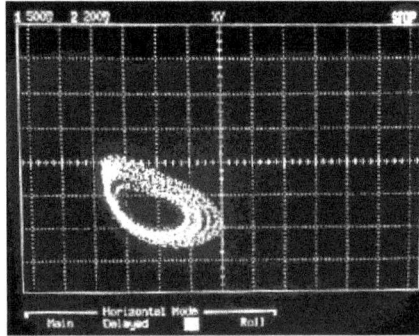

Fig. 7.9 Experimental results: projection of the single scroll strange attractor onto the $x - y$ phase plane.

Chua's circuits is taken into account. When the array consists of identical units, spatio-temporal chaos is evident, whereas introducing a spatial dissymmetry leads to an organized behavior and the emergence of spatio-temporal patterns.

Being implemented by CNNs, the cell introduced here is particularly suitable for carrying out experimental studies on these complex systems consisting of locally coupled units.

The integrated cell introduced in this chapter is important to provide an experimental framework for the study of these complex behaviors. The SC technique allows to obtain a VLSI system consisting of many cells. For instance, since the area consumption of a single cell is $200\mu m \times 200\mu m$, in a $2mm \times 2mm$ VLSI chip 100 cells can be integrated.

As an example of the complex behavior that can be obtained with our cell, the results of a simulation of three coupled Chua's circuits are shown. The dimensionless equations of the three circuits are:

$$
\begin{aligned}
\frac{dx_k}{dt} &= \alpha_k[y_k - h(x_k)] \\
\frac{dy_k}{dt} &= x_k - y_k + z_k + D_k(x_{k+1} - 2x_k + x_{k-1}) \\
\frac{dz_k}{dt} &= -\beta_k \cdot y_k \\
h(x_k) &= m_{1,k} \cdot x_k + 0.5 \cdot (m_{0,k} - m_{1,k}) \cdot [|x_k + 1| - |x_k - 1|]
\end{aligned}
\tag{7.1}
$$

where $k = 1, 2, 3$. The circuits have slightly different parameters as shown in Table 7.1, where the parameter D was introduced in order to change the diffusion coefficient. For sake of simplicity the RC circuit has been taken into account. Each circuit is chaotic when uncoupled. When coupled, the three circuits show "not in-phase" synchronization. Moreover, the synchronization pattern changes as parameter D is varied.

Table 7.1   Circuit parameter of the three cell array
described in Eq. (7.1).

	α_k	β_k	$m_{0,k}$	$m_{1,K}$	D_k
Circuit 1	9	15.50	-0.26	0.21	99D
Circuit 2	9	14.13	-0.20	0.25	94.5D
Circuit 3	9	12.82	-0.14	0.29	90D

Our experiment confirms the scenario presented in [De Castro *et al.* (1995)]. As the parameter D is varied, different synchronization patterns appear as shown in Fig. 7.10. A period-doubling scenario is evident. For $D=0.01$ the circuits synchronized each other with a phase lag visible in Fig. 7.10a. As the parameter D is decreased, 2T (Fig. 7.10b), 4T (not shown) and 8T (Fig. 7.10c) synchronization appear. For $D=0.0050$ the phases of the circuit state variables are synchronized, but their amplitudes are chaotic.

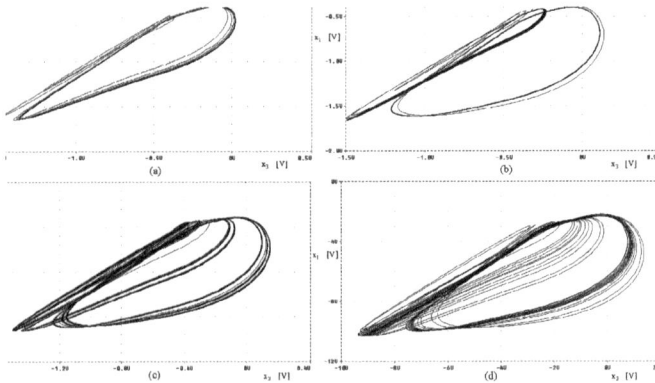

Fig. 7.10   SPICE simulation of an array of three coupled Chua's circuits. Variable x_1 versus x_3 is plotted. (a) $D= 0.01$, 1T synchronization; (b) $D= 0.0059$, 2T synchronization; (c) $D= 0.0054$, 8T synchronization; (d) $D= 0.0050$, Chaos.

The integrated Chua's circuit discussed in this Chapter represents a milestone in the research on chaotic circuit, because of the opportunity to experimentally investigate complex systems made of chaotic units.

Chapter 8

The four-element Chua's circuit

8.1 Model

The Chua's circuit consists of five elements: two capacitors, one inductor, a nonlinear resistor and a linear resistor. Four of these elements are without any doubt strictly necessary to design a circuit exhibiting chaos. The question about the need to have at least one further resistor always puzzled Chua and other scientists working on the Chua's circuit. In 2008 Barboza and Chua [Barboza and Chua (2008)] finally found an answer to this question: a Chua's circuit with only four elements can be designed.

The starting point for their analysis is one of the eight topologies initially hypothesized by Chua (see Chapter 1, Fig. 1.8(d)) as a candidate for generating chaos. The circuit is shown in Fig. 8.1. This topology is indeed not compatible with the constraints formulated by Chua: in fact, he required a $v_R - i_R$ characteristic confined in the second and in the fourth quadrant and such that the circuit had three equilibrium points. Barboza and Chua overcame this problem using a different three-segment nonlinearity (shown in Fig. 8.2) such that three equilibrium points can be effectively obtained for $i_R = 0$ (*i.e.*, when the nonlinear element is open circuit). In terms of the Chua's equations (1.7), this implies to implement $h(x)$ instead of $f(x)$, *i.e.*, a nonlinearity with only one negative slope. Barboza and Chua, then, demonstrated that the circuit is linearly equivalent to the Chua's circuit and that, furthermore, it exhibits a double scroll chaotic attractor even in the limit case of $R = 0$.

Let us first analyze the equations governing the circuit shown in Fig. 8.1. They can be easily derived by applying the Kirchhoff's circuit laws:

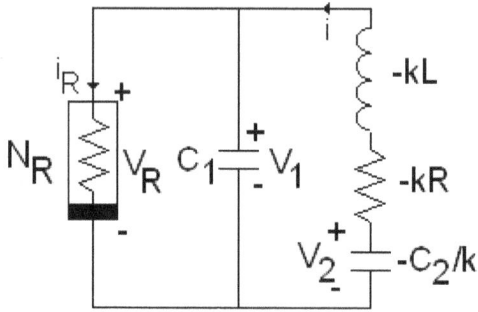

Fig. 8.1   A circuit linearly equivalent to the Chua's circuit.

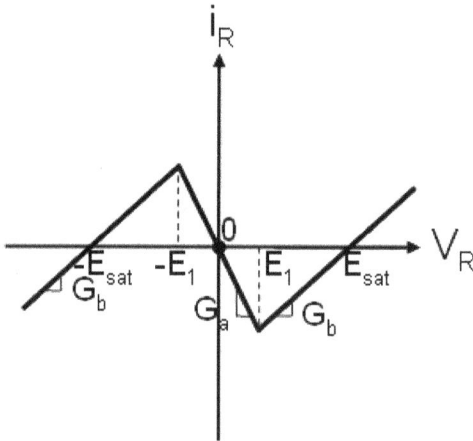

Fig. 8.2   Nonlinearity used in the circuit of Fig. 8.1.

$$\frac{dv_1}{dt} = \frac{1}{C_1}[i - g(v_1)]$$
$$\frac{di}{dt} = \frac{1}{\kappa L}[v_1 - \kappa R i - v_2] \qquad (8.1)$$
$$\frac{dv_2}{dt} = \frac{1}{C_2}\kappa i$$

where it should be noticed that the inductor current i now plays the role of the second state variable y of the Chua's equations (1.7), while v_2 that of the third state variable z. This can be explicitly demonstrated by deriving the dimensionless equations associated to this circuit. Taking into account the following rescaling:

$$
\begin{aligned}
x &= v_1/E_1 \\
y &= \sqrt{L/C_2}\,i/E_1 \\
z &= -v_2/E_1 \\
\tau &= t/\sqrt{LC_2} \\
\rho &= \frac{R}{\sqrt{L/C_2}} \\
\sigma &= C_2/C_1 \\
\kappa &= \frac{R_1}{R_2} \\
m_0 &= \sqrt{\tfrac{L}{C_2}}\,G_a \\
m_1 &= \sqrt{\tfrac{L}{C_2}}\,G_b
\end{aligned}
\tag{8.2}
$$

one obtains:

$$
\begin{aligned}
\frac{dx}{d\tau} &= \sigma[y - h(x)] \\
\frac{dy}{d\tau} &= \frac{x+z}{\kappa} - \rho y \\
\frac{dz}{d\tau} &= -\kappa y
\end{aligned}
\tag{8.3}
$$

with the usual definition of $h(x)$. With the further time rescaling $\tau' = \tau/\kappa$, equations (8.3) become:

$$
\begin{aligned}
\frac{dx}{d\tau'} &= k\sigma[y - h(x)] \\
\frac{dy}{d\tau'} &= x - \chi y + z \\
\frac{dz}{d\tau'} &= -\kappa^2 y
\end{aligned}
\tag{8.4}
$$

that match Chua's equations (1.7) for $\alpha = \kappa\sigma$, $\beta = \kappa^2$, $\chi = \kappa\rho = 1$. So, in the more general case, the circuit of Fig. 8.1 has a further bifurcation parameter, but if one fixes $\chi = 1$ its dimensionless equations (8.4) are identical to the Chua's equations (1.7).

The circuit shown in Fig. 8.1 is linearly conjugate to the Chua's circuit. This means that each dynamics exhibited by the Chua's circuit can be qualitatively reproduced by the circuit of Fig. 8.1 and viceversa. In particular, to map the behavior of the Chua's circuit into the circuit of Fig. 8.1, one can first calculate the parameters p_1, p_2, p_3, q_1, q_2 and q_3 from Eqs. (1.30) and (1.32) and then apply the following equations to derive the circuit components of the circuit of Fig. 8.1:

$$C_1 = 1$$
$$\frac{C_2}{\kappa} = 1 - \frac{p_1 + P}{p_3}(p_2 + P(p_1 + P))$$
$$\kappa L = \frac{p_1 + P}{p_3 - (p_1 + P)(p_2 + P(p_1 + P))}$$
$$\kappa R = \frac{P(p_1 + P)}{p_3 - (p_1 + P)(p_2 + P(p_1 + P))} \tag{8.5}$$
$$G_a = -(p_1 + P)$$
$$G_b = -(q_1 + P)$$

where $P = \frac{p_1 q_3 - p_3 q_1}{p_3 - q_3}$. These equations are analogous to Eqs. (1.38) that allow to derive the circuit parameters of the Chua's oscillator starting from a given set of parameters p_1, p_2, p_3, q_1, q_2 and q_3. Furthermore, analogously to Eqs. (1.38), also in Eqs. (8.5) C_1 is a free parameter. In this case, the parameter κ can also be arbitrarily chosen.

Equivalently, if one wants to derive the circuit parameters of the Chua's circuit corresponding to a given dynamics of the circuit of Fig. 8.1, first, the parameters p_1, p_2, p_3, q_1, q_2 and q_3 for the circuit of Fig. 8.1 should be derived according to:

$$\frac{G_a}{C_1} + \frac{R}{L} = -p_1$$
$$\frac{G_a R}{C_1 L} + \frac{1}{C_2 L} - \frac{1}{\kappa L C_1} = p_2 \tag{8.6}$$
$$\frac{G_a}{C_1 C_2 L} = -p_3$$

and

$$\frac{G_b}{C_1} + \frac{R}{L} = -q_1$$
$$\frac{G_b R}{C_1 L} + \frac{1}{C_2 L} - \frac{1}{\kappa L C_1} = q_2 \tag{8.7}$$
$$\frac{G_b}{C_1 C_2 L} = -q_3$$

Then, such parameters should be used to derive the circuit components to be used in the corresponding Chua's circuit as follows:

$$C_1 = 1$$
$$C_2 = \frac{p_3 - p_2(p_1 + P) - P(p_1 + P)^2}{P^2(p_1 + P)}$$
$$L = \frac{P^2(p_1 + P)^2}{p_3(p_3 - p_2(p_1 + P) - P(p_1 + P)^2)}$$
$$R = \frac{P(p_1 + P)}{p_3 - p_2(p_1 + P) - P(p_1 + P)^2} \tag{8.8}$$
$$G_a = \frac{p_2(p_1 + P) - p_3}{P(p_1 + P)}$$
$$G_b = \frac{q_2(p_1 + P) - p_3}{P(p_1 + P)}$$

where $P = \frac{p_1 q_3 - p_3 q_1}{p_3 - q_3}$.

The importance of the circuit of Fig. 8.1 is that a double scroll chaotic attractor can be generated even when $R = 0$. In this case the circuit reduces to the scheme shown in Fig. 8.3. This leads to the conclusion that only four components, namely three energy-storage elements (two capacitors and an inductor) and a nonlinear resistor are needed to design a chaotic circuit.

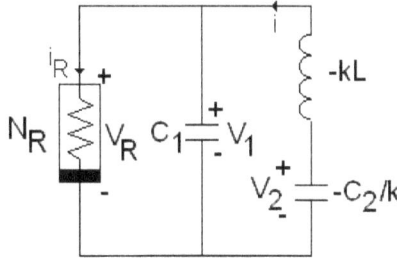

Fig. 8.3   The four-element Chua's circuit.

It is interesting to note that an exact mapping of the four-element Chua's circuit to the original Chua's circuit cannot be obtained. In fact, substituting $R = 0$ in Eqs. (8.6) and (8.7), one obtains $p_2 = q_2$ which does not satisfy the constraints discussed in Chapter 1. Only a perturbation of this parameter ($R \neq 0$) leads to a dynamics that can be implemented in the Chua's circuit.

8.2   Implementation

As discussed above, the four-element Chua's circuit consists of four elements. Since one capacitor and an inductor have negative values, the practical realization requires two operational amplifiers. One of them is needed to implement the nonlinearity (in this case it is in fact a three-segment piecewise linear function that can be realized using the scheme of Section 2.2). The second operational amplifier is needed to realize capacitor and inductor with negative values.

The complete scheme which refers to the more general case with $R \neq 0$ is shown in Fig. 8.4. Figures 8.5(a) and 8.5(b) show two attractors shown by this circuit when R is varied. The parameters of the circuit can be chosen according to [Barboza and Chua (2008)]: $C_1 = 33nF$, $C_2 = 100nF$, $L = 10mH$, $R_1 = 1k\Omega$, $R_1 = 8.2k\Omega$, $R_3 = 10k\Omega$, $R_4 = 9.8k\Omega$, $R_5 = 780\Omega$,

$R_6 = 1.2k\Omega$. Two operational amplifiers (for instance, integrated in a single device TL082) powered at $\pm 15V$ can be used. Figure 8.5(a) refers to the case $R = 211\Omega$ (single scroll attractor, while Fig. 8.5(b) to $R = 256\Omega$ (double scroll strange attractor).

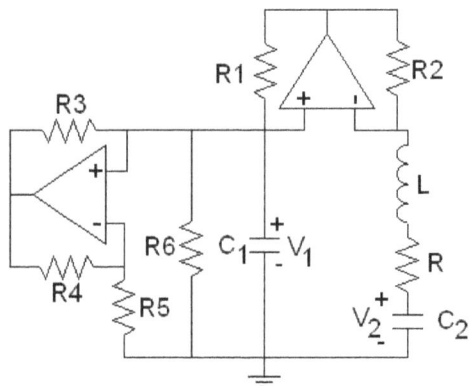

Fig. 8.4   Implementation of the four-element Chua's circuit using operational amplifiers.

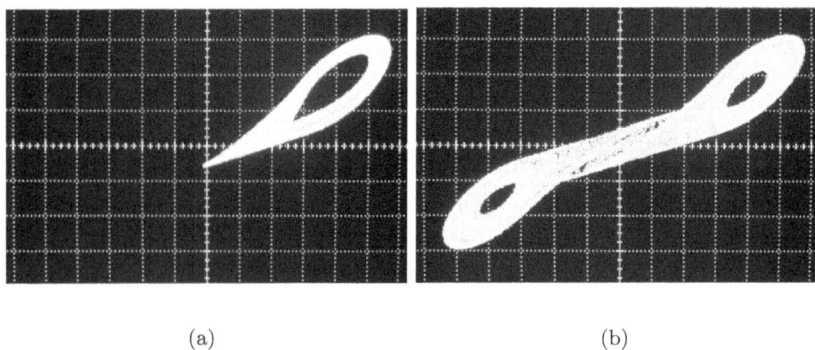

(a)   (b)

Fig. 8.5   Four-element Chua's circuit. (a) Projection on the plane $v_1 - v_2$ of the single scroll attractor. (b) Projection on the plane $v_1 - v_2$ of the double scroll attractor. Horizontal axis: $200mV/div$; vertical axis $500mV/div$.

8.3   Chua's circuit and the memristor

To conclude this Chapter, we would like to briefly mention that the research on the topological aspects connected to the Chua's circuit is now oriented towards the possibility of building a chaotic circuit, perhaps a Chua's circuit, by using *memristors*, i.e., the fourth canonical element. A research group at the University of California, Berkeley, is currently involved in this research project.

The existence of a fourth canonical element, called memristor, has been pointed out by Leon O. Chua in 1971 [Chua (1971)], but only recently the first physical example of a memristor appeared [Strukov *et al.* (2008)].

In his seminal work [Chua (1971)] Chua observed that there are four basic variables (the current i, the voltage v, the charge q, and the magnetic flux φ) describing a circuit and six different mathematical relations connecting two of these four variables. Two of these relations are determined by the definitions of two variables, the flux is the time integral of the electromotive force and the charge is the time integral of the current. Three of these relations describe the three well-known basic electrical elements, *i.e.*, the capacitor, the resistor and the inductor.

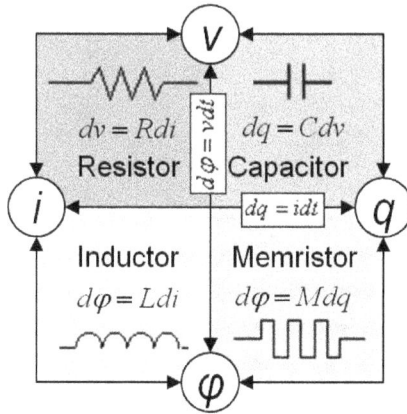

Fig. 8.6   The four electrical basic elements.

From these considerations, Chua argued that it must exist a fourth element, the memristor M, providing the last relation, *i.e.*, that between charge and flux, as shown in Fig. 8.6. In the case of linear elements, M is a constant and is identical to a resistance, but when M is a function of the charge q, it constitutes a nonlinear element which acting as a resistor with memory has a lot of potentially interesting applications.

Recently, a research group at the HP Labs [Strukov *et al.* (2008)] showed how memristance can effectively arise in a nanoscale system. The authors of this research [Strukov *et al.* (2008)] showed how a resistive system can be obtained by fabricating a layered platinum-titanium-oxide-platinum nanocell device.

The great interest generated by this discovery is evident if one observes the large number of commentaries and reports on this new element appeared on: Nature New, Scientific American, IEEE Spectrum, IEEE CAS Magazine, The New York Times, BBC News, The Wall Street Journal, Wikipedia, Washingtonpost.com and many others. In the future the memristor can play a fundamental role both as a model of phenomena arising in nanoscale electronics and as a key element to increase computing time over time [Tour and He (2008)].

Chapter 9

The organic Chua's circuit

In the last decade a wide interest has been directed at organic materials with semiconductor properties. The development of organic technologies represents for electronic engineers and material science experts an important research topic towards the development of innovative applications. In the post-silicon era, organic technology is born from the necessity to exploit more economical techniques for device processing and realization, in terms of reduced manufacturing and wafer costs. Current integrated circuit technology is extremely expensive due to the need to build ultra clean rooms allowing only a few contaminant molecules in the air (no more than one particle greater than $0.5 \mu m$ in one cubic foot of air). Organic devices can be realized in ordinary rooms with low-cost technology such as large production methodologies like reel-to-reel, inkjet printing, *etc* [Gamota *et al.* (2004)]. In addition, organic circuits can be printed on bendable substrates such as plastics and wearable garments.

The proposed technology [Fortuna *et al.* (2008)] is based on a new generation of active electronic devices: the Organic Thin-Film Transistors (OTFTs). This fundamental electronic device has been firstly reported in 1983 [Ebisawa *et al.* (1983)] and its performance has been continuously enhanced by investigating new materials and processing flow-charts [Dimitrakopoulos and Mascaro (2001)]. However, the reduced mobility of organic materials is the key point to be addressed in the development of this technology. Moreover, since, in the existing technology, n-type organic semiconductors have a much lower mobility and stability than p-type, the use of a complementary logic approach is not feasible. Thus, today most of the OTFT applications are limited to simple digital circuits, such as basic logic gates, ring oscillators and logic circuits [Hagen *et al.* (1999)]. A breakthrough in this field could be obtained by intense material scouting and by conceiving new circuit architectures for microelectronic system design.

To implement the Chua's circuit with organic transistors [Fortuna *et al.* (2007)], two main issues have to be addressed: the limits of the technology and the absence of elementary analog blocks based on OTFTs. The limits of the technology are mainly due to the low values of mobility and the small currents that imply slow dynamic response to applied signals. The absence of elementary analog blocks based on OTFT requires ad hoc electronic circuital solutions which have been developed in this work.

The Chua's circuit is here implemented by using electronic blocks that are based on only *p*-type organic transistors. Moreover, novel blocks have been conceived to implement analog systems that perform nonlinear dynamics.

9.1   The OTFT device

Organic thin-film transistors are field-effect devices with organic thin-film semiconductors as active layer. In terms of performance, organic transistors cannot be compared to inorganic semiconductor based transistors, however, the opportunity to realize low cost, large area, and flexible device by using low-cost manufacturing techniques can open the way to the development of other electronic applications [Gamota *et al.* (2004)]. To date, several studies have also been performed on *n*-type organic transistors but these transistors still remain not stable and feasible, unlike its *p*-type counterpart.

In this implementation, due to the limits of the technology only *p*-type transistors, with the top gate architecture illustrated in Fig. 9.1, have been used. The conduction channel between the drain and source contacts is characterized by the width W and the length L, and their ratio tunes the current flowing from the drain to the source as illustrated in the following.

The equations modeling the transistor behavior follow the well-known formulations based on the *Universal Mobility Law* (UML) and the *Variable Range Hopping* (VRH) model developed in [Brown *et al.* (1994)]. Thus, the OTFT, likewise silicon-based field effect devices, is characterized by three operating regions. These regions depend on the drain-source voltage V_{ds}, which should always have a negative value since the transistor is *p*-type.

The three regions differ from the expression of the drain current I_d as follows.

Depletion Region: the gate-source voltage V_{gs} is higher than the activation threshold voltage V_t. In this operating region, the drain current

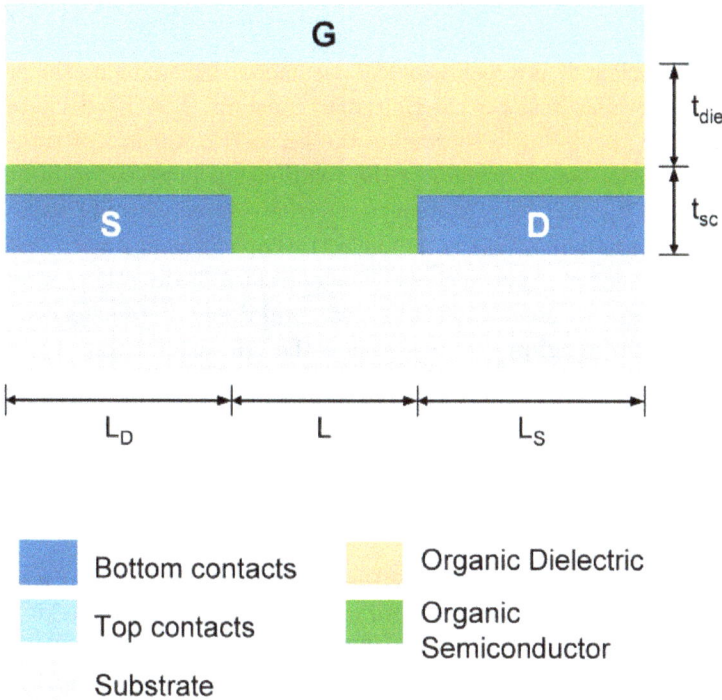

Fig. 9.1 Top gate architecture of the organic thin-film transistor.

can be considered constant and proportional to channel depth:

$$I_d = I_0 \cdot W \qquad (9.1)$$

Linear Region: if $V_{gs} < V_t$ the channel between drain and source contacts is created and the transistor is on. When drain-source voltage V_{ds} is higher than $V_{gs} - V_t$ the transistor works in the linear region and the drain current is described by the equation:

$$I_{d(linear)} = K \cdot \frac{W}{L} \cdot \left[|V_{gs} - V_t|^{2m+2} - |V_{gs} - V_t - V_{ds}|^{2m+2} \right] \qquad (9.2)$$

where:

$$K = \mu_0 \frac{1}{2m} \frac{C_{die}^{2m+1}}{(2m+1)(2m+2)(\varepsilon kT)^m}$$

The coefficient K is a technological parameter depending on the organic semiconductor features, k is the Boltzman constant, T is the absolute temperature and the $\frac{W}{L}$ ratio represents the geometric features of transistor. The exponential factor m models the relationship between drain current and voltages V_{gs} and V_{ds}. Moreover, there are three constant values: μ_0 represents the mobility, C_{die} is the dielectric capacity and $\varepsilon = \varepsilon_P \cdot \varepsilon_0$ is the organic semiconductor permittivity. The current I_d is directly proportional to the $\frac{W}{L}$ ratio that constitutes a fundamental design parameter.

Saturation Region: in this region the transistor is on ($V_{gs} \leq V_t$) and the drain-source voltage V_{ds} is lower than $V_{gs} - V_t$; the drain current expression is given by:

$$I_{d(sat)} = K \cdot \frac{W}{L} \cdot |V_{gs} - V_t|^{2m+2} \cdot (1 + \lambda \cdot V_{ds}) \tag{9.3}$$

where λ is a parameter quantifying the channel length modulation effect.

As it can be noticed, the transistor model is very similar to the MOSFET model, which is obtained for $m=0$. The term m, derived from the UML, takes into account the specific features of organic semiconductors.

The OTFT model parameters are extracted from the proprietary technology regarding materials and processing flows. Thus, the simulations that are presented below are based on real characteristics of existing OTFTs.

9.2 Implementation of the Organic Chua's Circuit

Starting from the above technological hypothesis the Chua's circuit has been designed by conceiving new electronic blocks based on only p-type OTFTs. The organic oscillator has been designed starting again from the CNN implementation, described by the following equations:

$$\begin{cases} \dot{x}_1 = -x_1 + a_1 y_1 + s_{11} x_1 + s_{12} x_2 \\ \dot{x}_2 = -x_2 + s_{21} x_1 + s_{23} x_3 \\ \dot{x}_3 = -x_3 + s_{32} x_2 + s_{33} x_3 \\ y_1 = 0.5 \cdot (|x_1 + 1| - |x_1 - 1|) \end{cases} \tag{9.4}$$

where y_1 is the piecewise linear (PWL) output function.

The system has been implemented by using the following values for the parameters : $a_1 = 27/7$, $s_{33} = s_{21} = s_{23} = 1$, $s_{11} = -11/7$, $s_{12} = 9$, $s_{32} = -14.286$.

Starting from equations (9.4), a rescaling has been introduced to match the OTFT technology features. The dimensionless state variables x_1, x_2, x_3 are modified according to suitable electrical values as in the following:

$$\begin{cases} x_1 = b_1 \cdot v_1 \\ x_2 = b_2 \cdot v_2 \\ x_3 = b_3 \cdot v_3 \end{cases} \tag{9.5}$$

Moreover, a linear relation between the CNN nonlinearity y_1 and its equivalent organic circuit implementation $i = g(v_1)$ is implemented by applying the following equation:

$$y_1 = A \cdot g(v_1) + B \tag{9.6}$$

Substituting equations (9.5) and (9.6) in Chua's state equations (9.4), the organic Chua's circuit equations are obtained as follows:

$$\begin{cases} C \cdot \dot{v}_1 = -g_{11}v_1 + g_{112}v_2 + A_g g(v_1) + I_1 \\ C \cdot \dot{v}_2 = -g_{22}v_2 + g_{21}v_1 + g_{23}v_3 + I_2 \\ C \cdot \dot{v}_3 = g_{32}v_2 + g_{33}v_3 + I_3 \end{cases} \tag{9.7}$$

A circuit implementing the nonlinear Chua's oscillator described by equations (9.7) is reported in the schematic of Fig. 9.2. The variables v_1, v_2, v_3 represents the voltages over capacitors C_1, C_2, C_3 with $C_1 = C_2 = C_3 = C$.

Silicon-based implementations of Chua's circuit are commonly realized by using *differential amplifiers* characterized by complementary integrated circuits.

According to the technological constraints, n-type transistors are not feasible for the implementation of the organic Chua's circuit. Thus, the design of blocks needed for the generation of the Chua's circuit dynamics requires new solutions which circumvent the use of n-type OTFT. Three new electronic architectures are designed using only p-type OTFT: a *saturation block* implementing the nonlinearity $g(v)$, a *transconductance amplifier block* and a circuit implementing a *negative current* contribution. The design of each block will be discussed in the following.

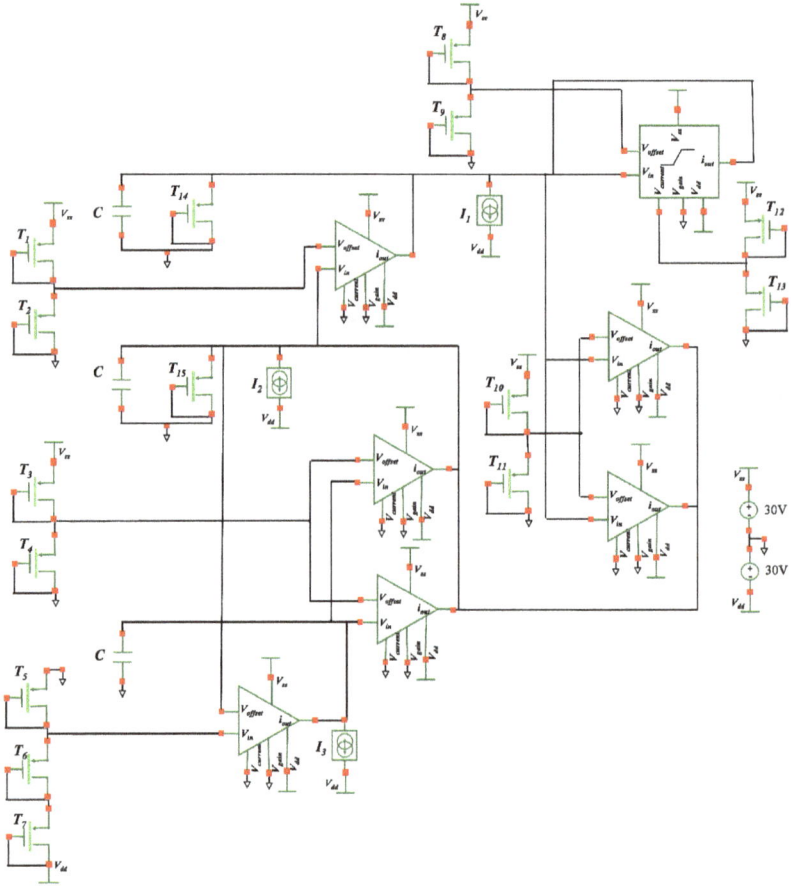

Fig. 9.2   Implementation of the organic Chua's circuit.

9.2.1   *Saturation Block*

The organic *saturation block* is designed to operate as a voltage-controlled current source. When the voltage input is within a specified range [v_{min}, v_{max}], the output current signal is proportional to the voltage input by the transconductance factor g_m. The output current saturates when the voltage input values are outside the specified range. The *saturation block* scheme, with only p-type OTFT, is shown in Fig. 9.3.

The *transconductance* value, the voltage and current ranges are totally configurable using appropriately chosen control inputs V_{offset}, V_{gain} and

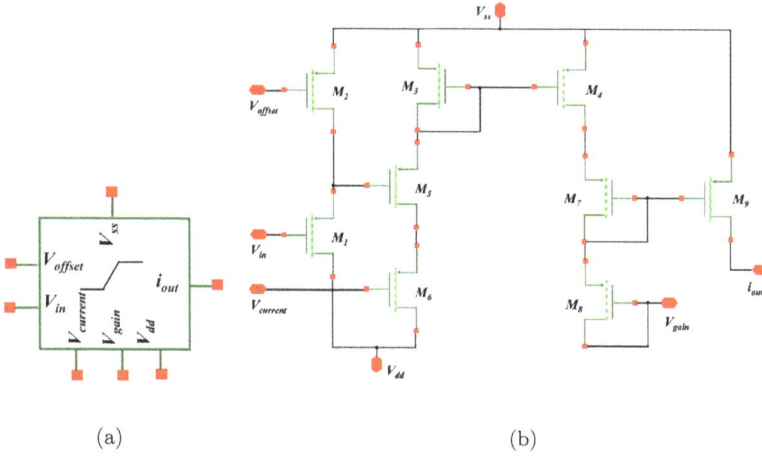

(a)   (b)

Fig. 9.3   Saturation block: (a) symbol; (b) circuit implementation.

$V_{current}$, as shown in Fig. 9.4. The g_m value can also be set up to obtain either positive or negative gains by swapping V_{offset} and V_{in} input terminals. The device is buffered so that the output current is independent of the voltage applied to the output port by external circuitry.

The *saturation block* is modeled by the following equation:

$$g(v_1) = \frac{I_M - I_m}{2} \cdot (1 + \frac{1}{v_{1max} - v_{1min}}) \cdot (|v_1 - v_{1min}| - |v_1 - v_{1max}|) + I_m$$

(9.8)

where I_m and I_M are respectively the minimum and maximum values of the output current and v_{min} and v_{max} are the breakdown points of the saturation function.

9.2.2   *Transconductance Amplifier Block*

The block is conceived to obtain a negative or positive configurable *transconductance amplifier*. According to the technology constraints, only field-effect transistors based on organic *p*-type semiconductors are used here. If the voltage input is within the specified range $[v_{min}, v_{max}]$, the input-output behavior is defined by the following relation:

$$I_{out} = g_m \cdot V_{in} + I_0$$

(9.9)

(a)   (b)

(c)

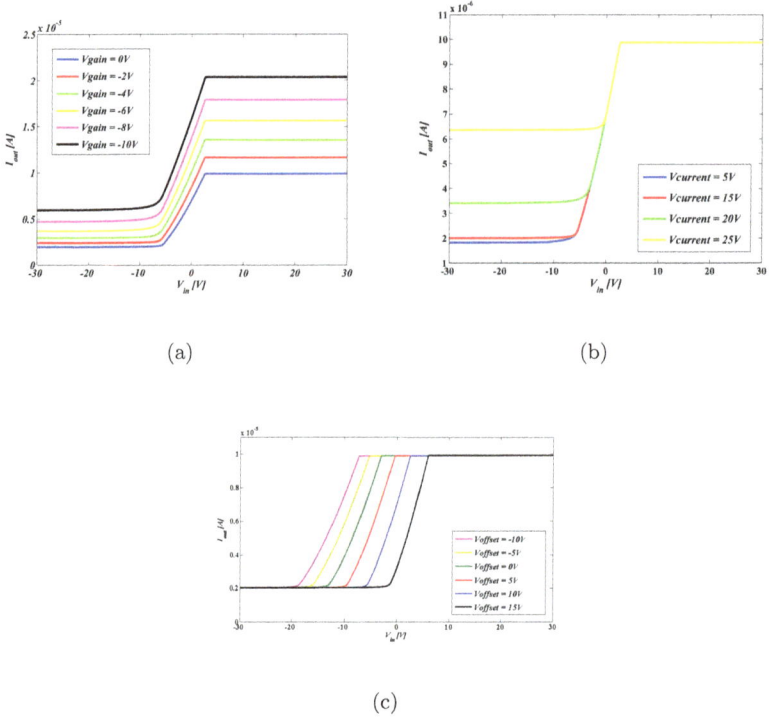

Fig. 9.4   Output current function dependence by: (a) V_{gain}, (b) $V_{current}$, (c) V_{offset}

Depending on the sign of g_m (negative or positive), two different circuit solutions have been adopted. The schematic symbol and the circuital implementation of the positive one are reported in Fig. 9.5. The *transconductance* value and the current ranges are set up by using appropriately chosen control inputs: V_{offset}, V_{gain} and $V_{current}$.

9.2.3   Negative Current Block

The main function of this organic block is to extract a constant negative current from a current node in the circuit. Generally, circuits performing such function use n-type current mirrors. Due to the adoption of only p-type OTFT, it is necessary to introduce a new circuital architecture exploiting a charge pump scheme as reported in Fig. 9.6. Moreover, the current value can be increased by adding few transistors between input and output nodes

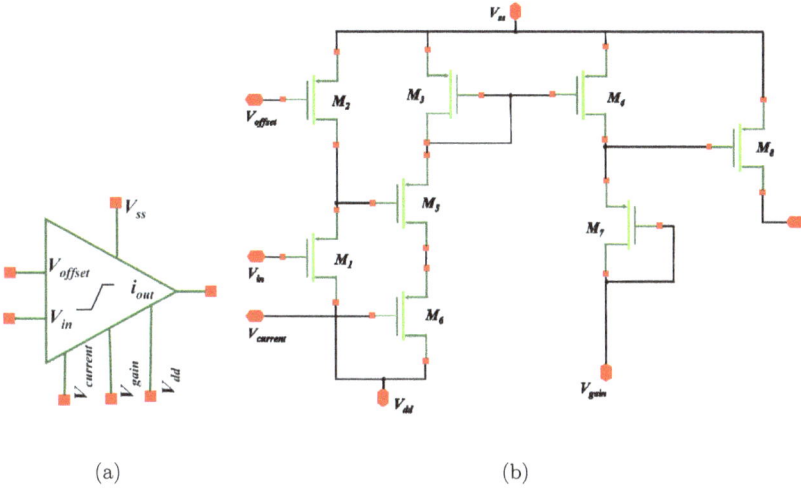

Fig. 9.5   Transconductance amplifier block: (a) symbol, (b) circuit implementation.

in the circuital scheme without increasing the architecture complexity.

9.3   Experimental Results

The organic Chua's circuit implementation is performed by rescaling Eqs. (9.5) and (9.6) with the parameters: $b_1 = 0.8333V^{-1}$; $b_2 = 0.1333V^{-1}$; $b_3 = 1.1667V^{-1}$; $A = 1.1494(1/\mu A)$; $B = -2.9540$.

The scheme shown in Fig. 9.2 has been implemented by using the organic blocks based on p-type OTFT and it has been simulated by using the *Eldo* and *Xelga* toolbox of *Menthor Graphics* (Cadence Unicad).

The system has been tested in order to reproduce the typical behaviors of Chua's oscillators. Figures 9.7 and 9.8 show the well-known double scroll chaotic attractor obtained from equations (9.7) by applying the following parameters: $C = 18.796 \ \mu F$; $g_{11} = 0.48333 \ \mu S$; $g_{12} = 0.27066 \ \mu S$; $g_{22} = 0.18796 \ \mu S$; $g_{21} = 1.1748 \ \mu S$; $g_{23} = 3.2893 \mu S$; $g_{32} = -0.05344 \ \mu S$; $g_{33} = 0 \mu S$; $A_g = 1$; $I_1 = -2.5332 \ \mu A$; $I_2 = 0 \ \mu A$; $I_3 = 0 \ \mu A$.

The geometric feature W of the transistors in the scheme of Fig. 9.2 are reported in Table 9.1.

By increasing the current I_1 in (9.7), the system switches to a single-scroll attractor as shown in Figs. 9.9 and 9.10 ($I_1 = -2.5760 \ \mu A$). Finally,

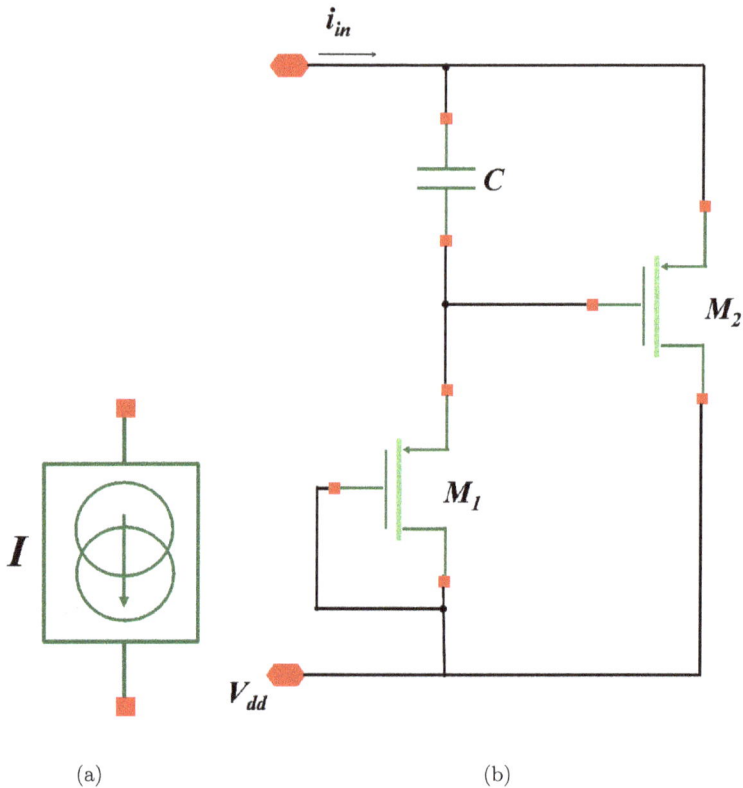

Fig. 9.6   Negative current block: (a) symbol; (b) circuit implementation.

a periodic behavior is obtained for $I_1 = -2.7497$ μA as shown in Figs. 9.11 and 9.12.

The results reported constitute the starting point in the development of new analog architectures based on organic materials paving the way to innovative applications in unexplored microelectronics fields. Although the dimension of the layout of the proposed scheme is much greater than the silicon-based implementations, the circuit could be realized at low cost for disposable electronic applications such as coded systems for product identification.

Table 9.1   Geometric features of designed transistors.

	W		**W**
T$_1$	4.2 mm	**T**$_9$	4 mm
T$_2$	10 mm	**T**$_{10}$	4.5 mm
T$_3$	3.4 mm	**T**$_{11}$	10 mm
T$_4$	10 mm	**T**$_{12}$	500 μm
T$_5$	500 μm	**T**$_{13}$	500 μm
T$_6$	500 μm	**T**$_{14}$	10.4 mm
T$_7$	500 μm	**T**$_{15}$	4.050 mm
T$_8$	425 μm		

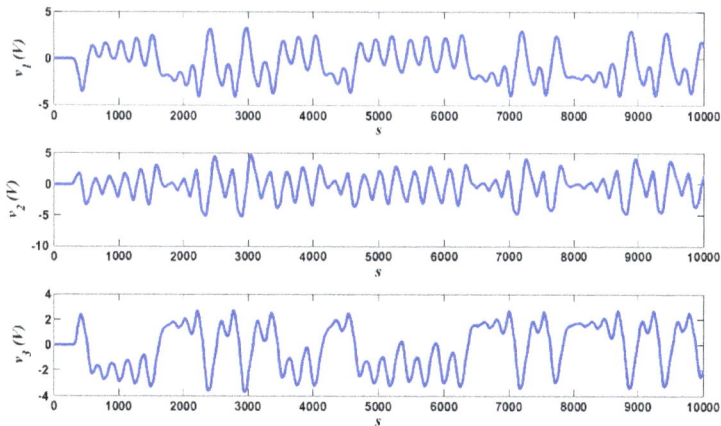

Fig. 9.7   State variables time-series of the organic Chua's oscillator for double-scroll chaotic dynamics ($I_1 =$-2.5332 μA).

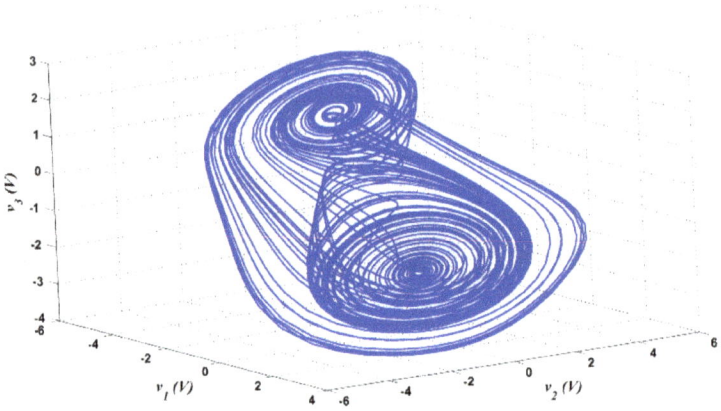

Fig. 9.8   Phase portraits of double-scroll chaotic dynamics from the organic Chua's oscillator (I_1 =-2.5332 μA).

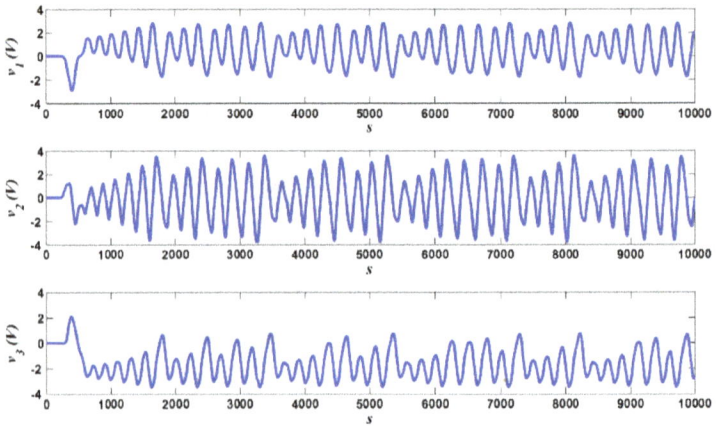

Fig. 9.9   State variables time-series of the organic Chua's oscillator for single-scroll chaotic dynamics (I_1 =-2.5760 μA).

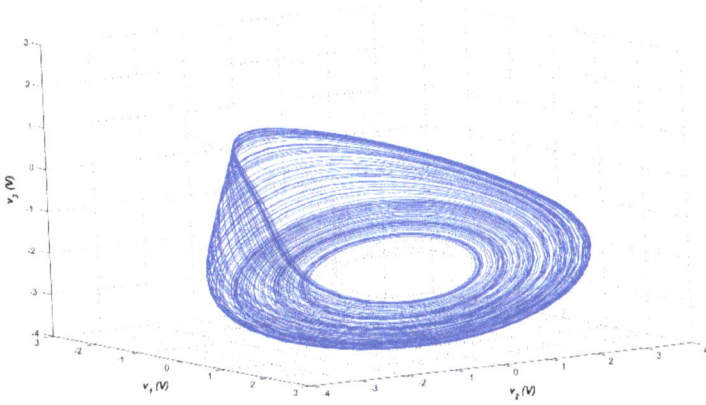

Fig. 9.10 Phase portraits of single-scroll chaotic dynamics from the organic Chua's oscillator (I_1 =-2.5760 μA).

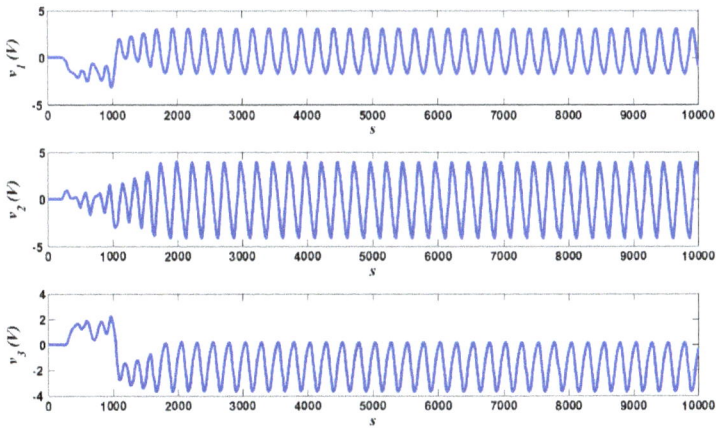

Fig. 9.11 State variables time-series of the organic Chua's oscillator for periodic dynamics (I_1= -2.7497 μA).

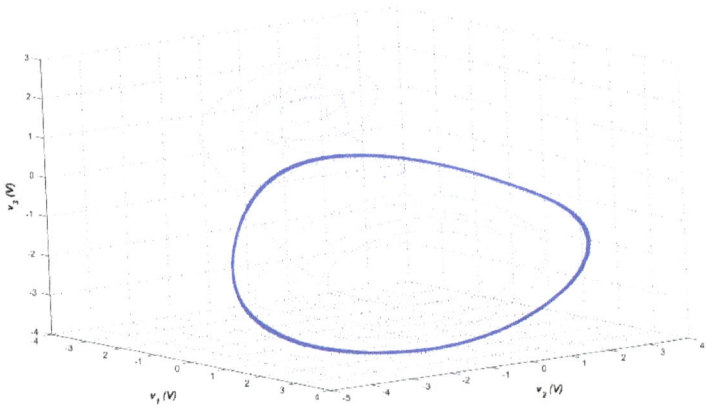

Fig. 9.12   Phase portraits of periodic dynamics from the organic Chua's oscillator ($I_1 =$ -2.7497 μA).

Chapter 10

Applications of the Chua's circuit

One of the most known applications of chaotic circuits is in the field of secure communications. As discussed in Chapter 1, several techniques for secure communications based on the use of a Chua's circuit have been developed. However, the Chua's circuit has been also applied to many other purposes. As an example, an approach for trajectory recognition based on Chua's circuit has been dealt with in [Altman (1993b)]. Another fascinating application of the Chua's circuit is obtained by designing a network of Chua's oscillators able of storing analogue patterns. Such a network can be used to face the problem of real-time handwritten character recognition [Baird *et al.* (1993)]. The Chua's circuit has been also applied to music with different music-making methods [Mayer-Kress *et al.* (1993); Rodet (1993a,b); Bilotta *et al.* (2005)]. We also mention that, starting from Bohr's model and the Chua's circuit bifurcation diagram the energy of electronic levels in atoms can be mapped to the parameter value α at which bifurcations occur, thus establishing a relationship between the bifurcation behavior of the circuit and the quantized energy level of atoms [Tonelli and Meloni (2002)]. Other applications have been developed for arrays of Chua's circuit, as discussed in Chapter 1.

Due to the low number of circuit elements, to the low cost of them and to the easy of implementation, the Chua's circuits has many important didactic applications in learning and experimenting chaos: many of the circuit implementations discussed in this book can indeed be implemented even at home! The authors of the book have been involved in several projects aimed at disseminating the Chua's circuit basic to young both undergraduate and high-school students.

In this Chapter we will discuss some recent applications in which chaos generated by a Chua's circuit is an essential feature to obtain high performance. The first application introduced in this Chapter consists in the realization and control of a walking microrobot actuated by piezoelectric materials [Buscarino *et al.* (2007b)]. Usually, the actuation of robot legs is controlled by square wave signals characterized by fixed amplitude and frequency. In this case however the control signals are generated performing a frequency modulation driven by the Chua's circuit. The smooth changes of the actuation signal frequency, performed by the chaotic system, enhances the microrobot walking capabilities especially when walking on given surfaces.

A second application concerns the use of the Chua's circuit as analog noise generator. This application is particularly important in the context of stochastic resonance and, in particular, in engineered systems in which the noise is intentionally added to achieve better performance. The Chua's circuit as analog noise generator is introduced in [Andò *et al.* (2000); Andò and Graziani (2000)] where it is demonstrated how it can be used to generate both Gaussian and uniform white noise.

A third application deals with the problem of improving performance of ultrasonic devices in presence of crosstalk and noise [Fortuna *et al.* (2003a)]. In order for each device to discriminate its own echo, chaos is exploited to create unique firing sequences. The firing scheme adopted is inspired to Chaotic Pulse Position Modulation (CPPM).

10.1   Chaos does help motion control

In this section we introduce an application of chaos to motion control in microrobotics. In particular, a microrobot actuated by piezoelectric elements, named PLIF (Piezo Light Intelligent Flea), designed to be fast, small, light and cheap, is taken into account [De Ambroggi *et al.* (1997); Muscato (2004); Buscarino *et al.* (2007b)] and chaos is used to enhance the motion capabilities on irregular surfaces. In fact, when driven with a constant frequency control signal, the microrobot is able to walk on regular surfaces if the frequency is appropriately tuned, but very small irregularities (such as grazes) can be a serious problem for the microrobot. By exploiting the widespread spectrum of a chaotic signal, a control signal with erratically varying frequency is provided to the robot making it able to deal with asperities in the surface and adaptable to different surfaces. In fact, in the considered microrobot chaos is directly used in the actuation system to modulate the signals devoted to the robot control.

Microrobotics is an emerging and constantly growing field of interest since the last decade. Micro and nano scale robots are surely useful in a wide range of application such as inspection of small environments, medicine, study of cooperating systems and so on [Dario *et al.* (1994)]. An important topic in microrobotics is the project of the actuation system that must accomplish low power consumption and small dimension specifics. In literature several microrobotic systems are described and they use different kind of actuators like pneumatic, electrostatic, shape-memory and piezoelectric [Conrad and Mills (1997)].

The actuation of the microrobot here adopted is based on piezoelectric ceramic actuators. Piezoelectric materials are particular structures able to produce a voltage when deformed and, viceversa, an excitation voltage induces a deformation that can generate a force. Hence, it is possible to use piezoelectric materials as deformation sensors as well as actuators. In Fig. 10.1(a) the scheme of a piezoelectric element used as an actuator is shown.

In this case two piezoceramics are joined and isolated through a resin coverage. Electrical contacts are realized by low temperature welding. The two elements are excited alternatively: the length of one of the elements, excited, is shortened while the other element stretches making the entire structure bending toward the short side. For the piezoelectric actuators used in this work, the displacement between the relaxed and the excited states is about 40 μm (Fig. 10.1(b)). To recover the original position it is sufficient to reverse the excitation voltage. The piezoelectric actuators are used to build the legs of the robot; each leg is therefore actuated by a flexor-extensor-like pair.

The whole structure of the PLIF robot, designed to be light and as small as possible, is shown in Fig. 10.1(c). We choose the simplest stable configuration using only three legs: the first one, placed in central-rear position, ensures the stability while the other two actually move the robot. Hence, the piezoceramic elements are not only the actuators but are also the physical support of the robot. In particular each leg is realized by two orthogonally linked piezoceramic elements, one performing the vertical movement (femur) and the other for the horizontal one (tibia).

10.1.1 *Motion control*

Aim of the motion control system is to generate and control the locomotion pattern of the microrobot. Each robot leg follows this movement sequence:

(a)

(b)   (c)

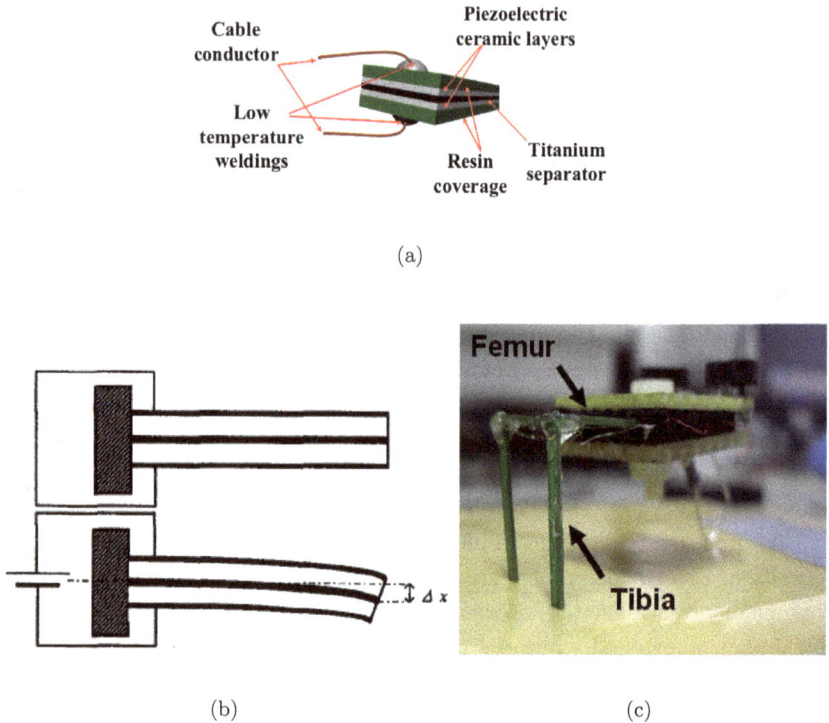

Fig. 10.1   (a) Piezoelectric scheme. (b) A piezoelectric actuator. (c) The PLIF micro-robot structure.

femur raising; tibia moving forward; femur going down; tibia moving backward. The motion pattern, which preliminary experimental tests have revealed to be the most effective for the adopted structure, is characterized by the simultaneous actuation of the two legs.

In [De Ambroggi *et al.* (1997); Muscato (2004)] and related works, the locomotion pattern is realized by an oscillator which generates a square wave signal with constant frequency and by a power circuitry (driver) providing the voltage supply needed by the piezoelectric actuator. To provide the robot with adaptive capabilities, this control scheme has been modified as shown in Fig. 10.2, where chaotic modulation of the control signal is introduced.

In this way, the generation of control signals with time-variant frequency can be accomplished. The frequency of the control signal changes as func-

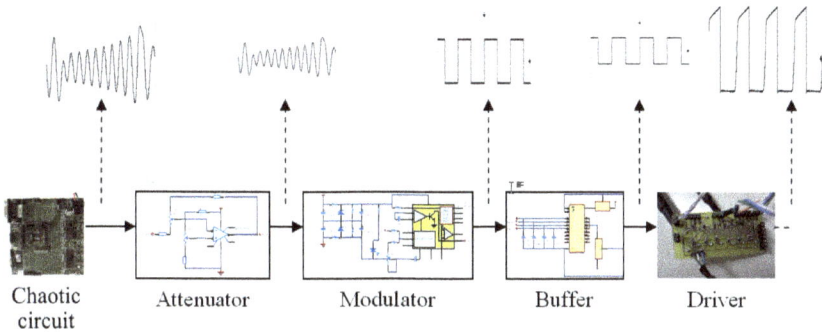

Fig. 10.2 Block scheme of the electronic board used to control the microrobot.

tion of the state variables of the Chua's circuit. Thus, the unpredictable behavior of the chaotic modulating signal is exploited to obtain a control system able to explore at each step new solutions to the motion control problem. In particular, one of the state variables of a Chua's circuit is used as modulation driving signal. Analyzing the spectrum properties of the three state variables, we have chosen the variable y to drive the modulation of control signals. In this way a signal with variable frequency in the range of 30-100 Hz can be obtained. The Chua's circuit is implemented on a FPAA board to dynamically change circuit parameters as described in Chapter 6.

The first block of the scheme of Fig. 10.2 is the Chua's circuit implemented by a FPAA board. The signal produced by this Chua's circuit oscillates with a peak-to-peak amplitude of about $1.1V$, while the modulator accept signals with amplitude not greater than $200mV$. Thus, an attenuation block, realized using an inverting operational amplifier, was introduced. The frequency modulator block is implemented using a XR2206, an oscillator able to generate square waves with frequencies from $1Hz$ to $1MHz$. Thus, the attenuated signal drives the generation of the variable frequency square wave signal which is then buffered and connected to the driver. In Fig. 10.3 it is possible to notice how the mean frequency of the control signal changes according to the dynamics of Chua's circuit. In fact, the chaotic modulation works on the frequency of the control signal and the resulting control signal is a square wave signal with fixed amplitude and variable frequency. The mean value of the control signal frequency depends on the value of the chaotic variable used for the modulation. Obviously, each leg needs two control signals, one for the femur and another one for the tibia. These two signals are the same but one is simply ninety degrees phase-shifted in order to obtain the correct movement of the leg.

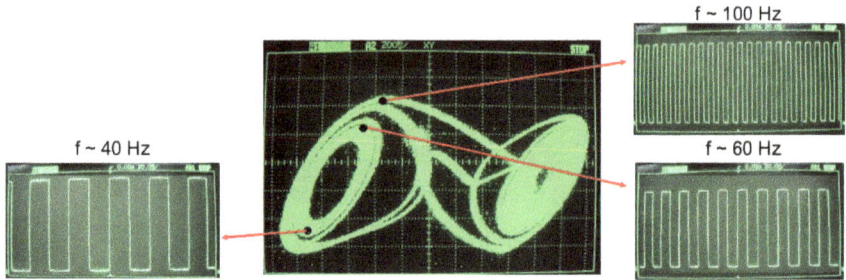

Fig. 10.3   Modulation of the control signal by the chaotic circuit.

10.1.2   *Experimental results*

In order to show the performances of the microrobot, we report three kinds of tests. In the first set of tests the microrobot walks on different smooth surfaces like an iron or wooden layer. In the second set of tests the surfaces are grazed in order to compare the performances in terms of velocity obtained with chaotically modulated control signals and with constant frequency control signals. The last set of tests concerns the overloading of the microrobot structure in order to verify if the introduction of chaotically modulated control signals is able to improve the motion also in presence of heavy structures.

Comparative results show that driving the actuation by using chaotic modulation leads to consistent improvements in terms of two factors. The first advantage is difficult to be quantified: often on grazed surfaces the robot without chaotic control stops when it encounters some surface asperity. The second advantage is in terms of robot speed as illustrated in Table 10.1, which reports the measures of the time the robot needed to walk on a given path for a fixed length on an iron layer. The constant frequency case is performed using a square wave signal with a frequency of $82 Hz$: this value corresponds to the best case obtained in the preliminary tests. For each surface five tests have been performed and mean value and variance of the data have been reported.

For regular surfaces, the average velocity increases when using the chaotic approach. Similar results have been observed grazing the test surfaces. In these tests, the robot, driven by chaotically modulated signals, is able to pass over the scratches while in the case of constant frequency actuation signal the robot often stops or decreases its velocity.

The last set of tests are performed overloading the robot structure. Results, loading the robot with a load about ten times heavier than the original weight, show that using our approach the robot is able to slowly move although it was completely stopped when actuated using constant frequency signals. Results clearly show that while the robot is not able to complete the path in the case of constant frequency control, it is able to slowly moves when controlled by chaotically modulated frequency.

To graphically show the improvements obtained using chaotically modulated frequency signal, we equipped our robot with a led that lights up when the robot is actuated. We then use a camera with a long exposure time in order to take pictures which traces the robot trajectory. As shown in Fig. 10.4, improvements are clearly visible simply comparing the length of the red line.

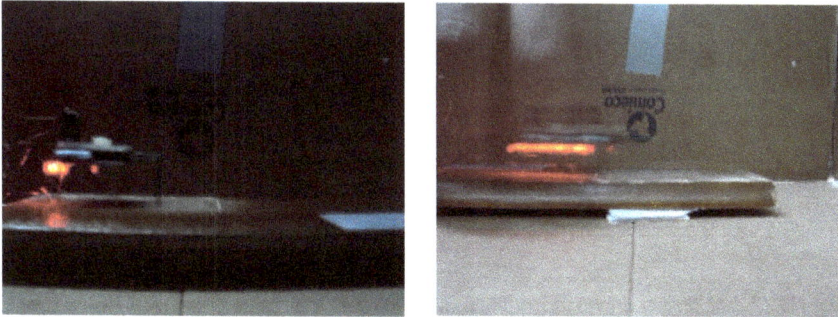

Fig. 10.4 Performances of the microrobot on an scratched wooden surface. On the left side the constant frequency case is reported, on the right side the chaotically modulated frequency case.

Table 10.1   Performances of the microrobot on different surfaces with and without chaos control.

		Mean value	Variance
		IRON LAYER	
Without chaos control	Time (s)	16.58	0.43
	Velocity (cm/s)	0.3	0.01
With chaos control	Time (s)	13.78	1.55
	Velocity (cm/s)	0.37	0.04
		GRAZED IRON LAYER	
Without chaos control	Time (s)	9.32	4.16
	Velocity (cm/s)	0.21	0.14
With chaos control	Time (s)	7.58	1.6
	Velocity (cm/s)	0.27	0.06
		OVERLOADED IRON LAYER	
Without chaos control	Time (s)	over	-
	Velocity (cm/s)	0	0
With chaos control	Time (s)	28	3.9
	Velocity (cm/s)	0.072	0.009
		WOODEN LAYER	
Without chaos control	Time (s)	17.96	0.54
	Velocity (cm/s)	0.39	0.01
With chaos control	Time (s)	10.9	1.39
	Velocity (cm/s)	0.65	0.08
		GRAZED WOODEN LAYER	
Without chaos control	Time (s)	9.12	3.31
	Velocity (cm/s)	0.18	0.11
With chaos control	Time (s)	7.28	1.1
	Velocity (cm/s)	0.28	0.04
		OVERLOADED IRON LAYER	
Without chaos control	Time (s)	over	-
	Velocity (cm/s)	0	0
With chaos control	Time (s)	32	2.1
	Velocity (cm/s)	0.061	0.009

10.2   The Chua's circuit as an analog noise generator

The idea that noise usually degrades the performance of a system has been overcome by the concept of stochastic resonance [Benzi *et al.* (1981); Gammaitoni *et al.* (1998)]: in nonlinear dynamical systems noise can have benefic effects. Now the concept of stochastic resonance has been extended to account for all the cases in which the presence of noise enhances the degree of order of a system or improve its performance [Anishchenko *et al.* (2003)]. Examples are the motion of an overdamped Brownian particle in a bistable potential with weak periodic modulation, nonlinear circuits, lasers, biological systems. If noise can enhance the performance of an electronic circuit or a measuring device, one can think to intentionally add a noise source. For this reason, the availability of reliable techniques for analog noise generation becomes important.

   In [Andò *et al.* (2000); Andò and Graziani (2000)] such analog noise generator is obtained by properly configuring a Chua's circuit implemented with the CNN approach described in Chapter 3. More in details, Andò and Graziani demonstrated that the Chua's circuit can be used to generate either a Gaussian or an uniform white noise. To do this, they investigate the statistical and spectral characteristics of the signals generated by a Chua's circuit with respect to different values of the parameters α and β. They then applied the χ^2 method [Papoulis (1991)] to determine which set of parameters leads to a signal having the statistical and spectral characteristics more similar to those of either a Gaussian or uniform noise. By using this method, a signal (in particular, the state variable y) with a Gaussian-like distribution is obtained with a confidence of 95% for this set of parameters: $m_0 = -1/7$; $m_1 = 2/7$, $\alpha = 9.6$ and $\beta = 15.2$. Analogously, the same method allows to conclude that the set of parameters for which the Chua's circuit can be used as an analog generator of uniform white noise with a confidence of 95% is given by: $m_0 = -1/7$; $m_1 = 2/7$, $\alpha = 9$ and $\beta = 14.3$. The probability density functions corresponding to these two cases are reported in Fig. 10.5.

10.3   Chaotic Pulse Position Modulation to improve the efficiency of Sonar Sensors

Collision avoidance is one of the main issues involved when dealing with the design and construction of autonomous robots. Autonomous robot control

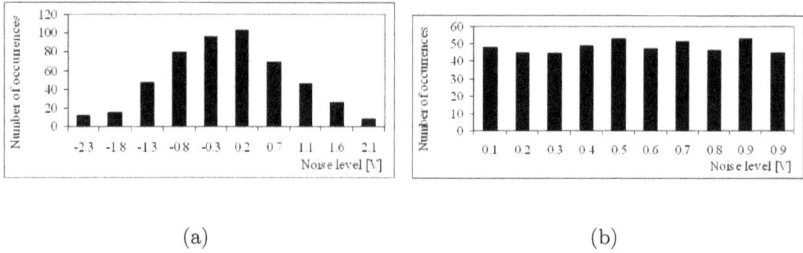

(a)   (b)

Fig. 10.5   Probability density function of (a) Gaussian-like signal or (b) uniform-like signal generated from a Chua's circuit.

often implies the ability of free-roaming platforms to travel in unstructured environments. The intelligent control of robots involves strategies to avoid obstacles and, if necessary, choose alternative routes to accomplish a task. This activity is performed on the basis of the information collected from several sensors. Among the most common sensors used for collision avoidance, those based on the measurement of the *Time of Flight* (TOF) are perhaps the most diffused. In particular, *sonars*, based on *Ultrasonic TOF ranging*, allow the user to conjugate reliability and precision of measurement, with low cost and ease of interface, making these sensors a widely used tool for autonomous robot navigation [Everett (1995)]. Autonomous robots are often equipped with a large number of ultrasound sensors to span the space around them and sense the presence of obstacles in their operational space.

Much work has been devoted to improve sonar performance, especially concerning target classification [Kleeman and Kuc (1994)], [Kleeman and Kuc (1995)], [Politis and Probert (1998)]. This is a task of primary interest for robot navigation, in which a map of an unstructured environment from known environment features is reconstructed. A classification standard for indoor target types in planes, corners, and edges is widely adopted in applications. A common strategy to enhance sonar performance is to use arrays of sensors, together with *integral* detection strategies (opposed to *spot* approaches). In particular in [Kleeman and Kuc (1995)], a minimal configuration of two pairs of transmitter-receiver has been adopted to classify targets thanks to a mathematical model of sound propagation and the storage of a set of echo templates. In [Politis and Probert (1998)], Politis and Probert deal with the definition of a frequency modulation of the operating frequency of the sonar, attempting to distinguish echoes from complex objects.

In the multi-user context, the phenomenon of *crosstalk* may give rise to frequent misreadings. Sensors in fact randomly influence each other, especially when a fast firing strategy is adopted for collision avoidance, leading to false range readings. Moreover, more and more robot applications require more than one robot to roam in the same environment. A practical expedient is to allow sensors to operate separately in time. More effective solutions to the problem of crosstalk involve the association of a unique 'signature' to each sensor, in order to detect own echoes and discard the others, allowing sensors to operate simultaneously. Crosstalk avoidance has been recently addressed in [Jörg and Berg (1998)], [Kleeman (1999)], [Borenstein and Koren (1994)]. The pioneering work [Borenstein and Koren (1994)] exploits the method of alternating delays in a traditional scheduled firing scheme to reduce the effect of noise and crosstalk. In [Kleeman (1999)], each sensor is driven by the emission of a couple of precisely timed pulses which uniquely identifies the transmitter.

Another interesting approach is the idea of applying pseudo-random codes as transmission sequence of the sonar [Jörg and Berg (1998)]. A suitable choice of pseudo-random codes to drive sonars can guarantee that the autocorrelation function of the sequence emitted is sharp and that two sequences coming from different sensors are uncorrelated. This is an ideal condition for the successful application of a detection strategy based on a matched filter technique [Woodward (1964)].

All these strategies need a computational unit: in [Borenstein and Koren (1994)] alternated delays should be generated and scheduled for each sensor of the array; in [Kleeman (1999)] a precise interval between pulses should be respected; whereas in [Jörg and Berg (1998)] a pseudo-random sequence should be computed.

In the following sections the peculiarity of chaos to enhance the performance of sonar systems in terms of crosstalk and noise rejection are exploited. The main idea underlying this approach is to drive a sonar with suitable chaotic signals and apply a matched filter technique for a robust rejection of crosstalk and noise.

10.3.1 *Continuous Chaotic Pulse Position Modulation*

The approach shown in this section conjugates the ideas of exploiting both the time intervals between the pulses emitted by a sonar and the peculiar characteristics of chaotic signals to build a unique signature belonging to each sensor. In particular, the strategy presented in this work uses the principles underlying the *Chaotic Pulse Position Modulation (CPPM)*[Maggio *et al.* (1999)], which has been introduced in the field of chaotic communications. The main idea underlying CPPM is to generate a sequence of pulses in which the duration of the time interval between a pulse and the next one is provided by a chaotic law. As the information is contained on the temporal distance between pulses, additive noise on the channel does not affect the integrity of the information. Moreover, pulses with a small duty cycle are used, thus involving low power consumption. The original formulation of CPPM is based on the generation of chaotic sequences by discrete maps, which require a digital unit. In our application a CPPM system in which the chaotic sequence is generated on the basis of a continuous-time chaotic attractor, rather than a chaotic discrete map is designed. This modification allows us to generate an effective chaotic sequence in an entirely analog fashion, without requiring any digital processor with a word of bits large enough to generate numbers with the required precision.

The original approach, based on discrete maps, allows us to assume a direct correspondence between the chaotic sequence x_k and the sequence of the time intervals τ_k between two pulses as follows:

$$x_k \longrightarrow \tau_k \qquad (10.1)$$

where k denotes the discrete time index. When, as in our case, a continuous chaotic circuit is used as chaotic generator instead of a discrete map, the modulation scheme has to be adequately modified in order to generate the sequence τ_k in Eq. (10.1).

In particular, the continuous CPPM has been essentially realized via a voltage-to-time conversion. This requires a circuit that is nonlinear and has its own dynamics which cannot be a priori neglected. Nevertheless, in the following it will be shown that a continuous CPPM can be realized with a good performance with respect to our goal, according to the preliminary work [Fortuna *et al.* (2001b)]. Let us consider in more details the continuous modulator, which is depicted in Fig. 10.6 by a block scheme. The chaotic sequence is generated as follows.

For sake of simplicity, we focus attention on a single state variable. A sample-and-hold (S/H) circuit perform a S/H operation on the continuous signal, which allows to get the value x_k. Subsequently, a ramp is generated and is stopped when its value equals that of the sampled signal x_k. When this occurs, a pulse is emitted from the modulator, the ramp is reset and a further S/H operation is performed. Therefore, the ramp signal reaches

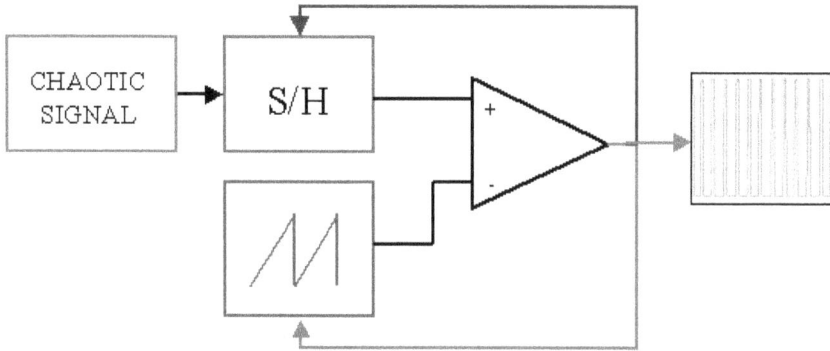

Fig. 10.6   Block scheme of the continuous chaotic pulse position modulator.

the value of the sample of the chaotic attractor x_k after a time τ_k which is proportional to the value x_k.

The circuit implementing such a modulator is shown in Fig. 10.7.

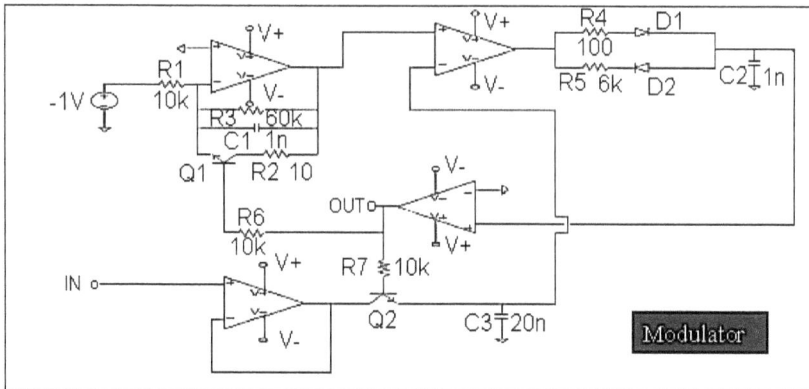

Fig. 10.7   Circuit implementation of the continuous chaotic pulse position modulator.

In our case, the analog circuit generating chaos is the Chua's Circuit. The modulator works only for positive values of the continuous chaotic signal. The introduction of a suitable offset is therefore required to guarantee the generation of only positive values.

In order to verify that the continuous CPPM approach allows us to preserve chaos, the pulse-coded continuous signal has been demodulated

and compared with the original one. The demodulation has been realized
by tracking the duty cycle of the modulated signal. The results illustrated
in Fig. 10.8 denote an evident correspondence between both signals and
their respective Fast Fourier Transforms.

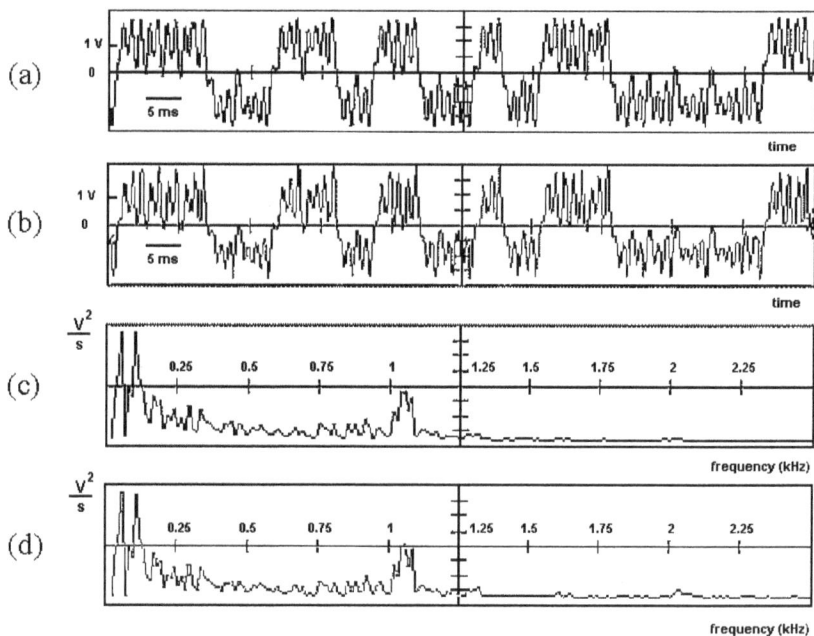

Fig. 10.8   (a) Trend of the state variable. (b) The demodulated signal. (c) Fast Fourier
Transform (FFT) of the signal in (a). (d) FFT of the signal in (b).

Furthermore, more in-depth comparisons between the two signals have
been made in order to check the adequacy of the modulation strategy,
leading to the conclusion that the information is not lost in the whole
sampling and modulation process.

10.3.2   The CPPM Sonar

Ranging measurements based on sonars are usually performed by measuring
the *TOF* of an ultrasound wave propagating in air. In low cost, autonomous
robot applications, the most used sonar sensor is perhaps the *Polaroid*

Series 600, driven by the *Polaroid 6500 Ranging Module* [Polaroid (1990)]. The ranging module provides all the needed circuitry to use the sensor in a simple fashion. In particular, the user has just to provide a INIT signal when a measurement has to be started. Subsequently, the ranging module allows the sensor to emit a train of 16 sound pulses at a frequency of 50kHz. Then, the module remains in a *WAIT* state until the reflected 16 pulses come back. Finally, an ECHO signal is raised to indicate the end of the measurement process. The user can retrieve the distance d by measuring the time $\bar{t}$ elapsed from the INIT to the ECHO, by performing the simple operation:

$$d = \frac{c\bar{t}}{2} \tag{10.2}$$

where c is the speed of propagation of the sound in air. This operation mode denotes strong drawbacks in the case of multi-user scenarios. It is worth remarking that even a single vehicle is usually equipped with more than one sensor in order to span the operational space over a large angle, which make almost any case a multi-user scenario. Here the sensor is driven by a continuous CPPM modulator, like the one depicted in Fig. 10.7. The output of the modulator is fed as input to the circuit illustrated in Fig. 10.9, in order to provide a suitable value of 400 V of amplitude to drive the sensor membrane. This circuit corresponds exactly to the power stage of the ranging module [Polaroid (1990)].

The circuitry implemented makes the sensor able to transmit and receive at the same time. This condition is different from the original operating condition of the sensor, in which it must be inhibited for 238ms in order not to receive the transmitted signal. It has been already proved that the maximum likelihood estimate of TOF of an echo pulse with additive Gaussian white noise is obtained by finding the maximum of the cross correlation function of the received pulse with the known pulse shape [Woodward (1964)]. Fundamental properties of cross correlation of chaotic signals ensure that the cross correlation between the transmitted and the received CPPM signals will show a significant peak in correspondence of the TOF. Next Section will show experimental results.

10.3.3   *Experimental results*

The experimental setup has been built with a Polaroid series 600 sensor. The power circuitry has been obtained by using only the output stage of the Polaroid series 6500 ranging module and inhibiting the remainder

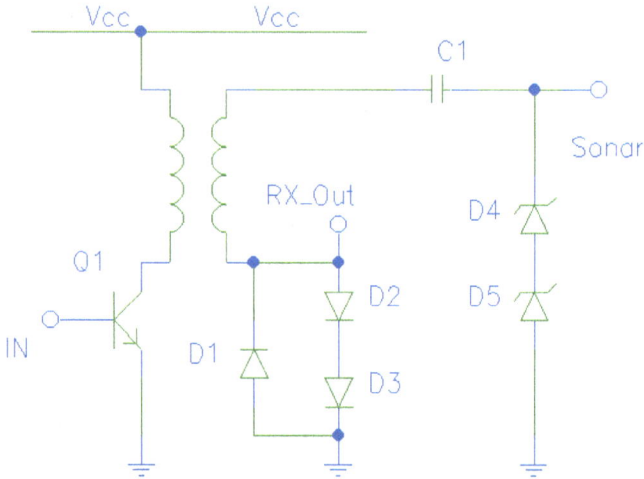

Fig. 10.9   Power stage of the CPPM driving circuit.

of the board. The train of pulses has been emitted through continuous CPPM driven by a Chua's circuit evolving according to a double scroll Chua attractor.

As a first example, experimental results related to a single measurement experiment are reported in Fig. 10.10.

It refers to a measurement of the distance of a target placed at 75cm. Fig. 10.10(a) reports the received signal in a time window of 200ms of duration, which coincides with the window used for performing the cross correlation between the transmitted and the received signals. The received signal is characterised by three contributions: the first one consists of the largest peaks, corresponding to the transmitted peaks, also received by the sensor, not inhibited during transmission. The smaller peaks are related to two sources: the first one is the real reflected ultrasound wave, while the second one is a reflected ultrasound wave coming from another sensor driven by CPPM relying on another double scroll Chua attractor with different parameters. This last source has been introduced to evaluate the performance of the sensor in a two-user scenario. Fig. 10.10(b) reports the cross correlation between transmitted and received signals. It shows two large peaks. One of them is centered at the sample no.20000, corresponding to origin of time; this is due to the fact that the transmitted signal is

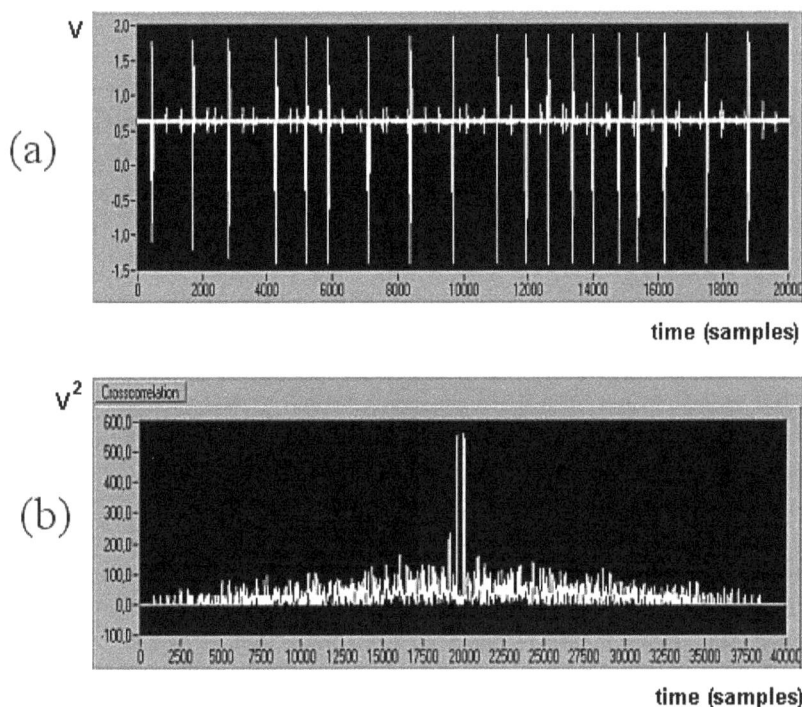

Fig. 10.10 Experiment with a target at 75cm. (a) Received signal. (b) Cross correlation between the transmitted and the received signals.

entirely enclosed in the received one. The other peak is centered at a time correspondent to the delay due to the TOF, correspondent to a distance of 73.48cm. Other smaller peaks are present, due to the noisy, multi-user condition.

The time window width has been chosen experimentally on the basis of the consideration that, in a matched filter strategy, the "time-bandwidth" product must be large enough. As the bandwidth of the sensor is physically limited, the only parameter on which one can act is the time window width. Unlike in [Jörg and Berg (1998)], keeping the time window width as small as possible (to guarantee a good autocorrelation function) is not needed. Chaotic signals in fact theoretically ensure a sharp autocorrelation function over time windows of any width.

The sensor has been characterised for measurements ranging from 5cm to 145cm, in 5cm steps. Three sets of experiments have been carried out. The first one refers to measurements performed by using the sonar driven by CPPM, in a single-user scenario. The second one refers to the same experiment, in presence of another CPPM sensor located close to the sensor to be characterised (two-user scenario). The third one refers to measurements performed by a sensor driven by an original Polaroid 6500-Series Sonar Ranging Module. It is worth remarking that in this case, measurements of distance under 40cm require specific techniques to damp the echo of the transmitted signal, which would lead to incorrect measurement. This is prevented by the constructor by introducing a blank interval of 238ms, when the sensor is inhibited. Fig. 10.11(a) and (b) reports the error committed and the measured distance, respectively, versus the real distance. Table 10.2 reports the average measurement error committed in the three cases.

To make a comparison, the table has been worked out by considering the range 40cm–145cm. It is worth noticing that the CPPM approach allows to perform measurements under 40cm without adopting any particular technique to damp the echo of the transmitted signal. The signal transmitted is in fact entirely received by the sensor and gives rise to a large peak of correlation at the origin of time, which can be ignored. In conclusion, the CPPM approach allows us to perform ranging measurements with an error comparable with that committed by the Polaroid Ranging module, despite of the presence of crosstalk and noise, thus obtaining a better overall performance.

Two important issues concern accuracy and Doppler effect. Firstly, the experimental results presented aim to show the effectiveness of chaos. For this reason, a very simple scheme has been adopted. Obviously, the measurement system performance can be improved by suitable strategies, like adopting more accurate propagation models, as a correction factor for the speed of sound [Kleeman and Kuc (1995)]. Secondly, like all integral methods, the CPPM sonar suffers from Doppler effect. When sensors are carried on a mobile robot, the effect of robot velocity gives rise to a Doppler shift in the pulse separation. In this approach, performance does not deteriorate for robot velocity up to 1.5 m/s.

Table 10.2   Average error committed in the range 40cm–145cm.

	Polaroid Ranging Module	CPPM	Multi-user CPPM
Mean	2.29%	1.87%	1.84%

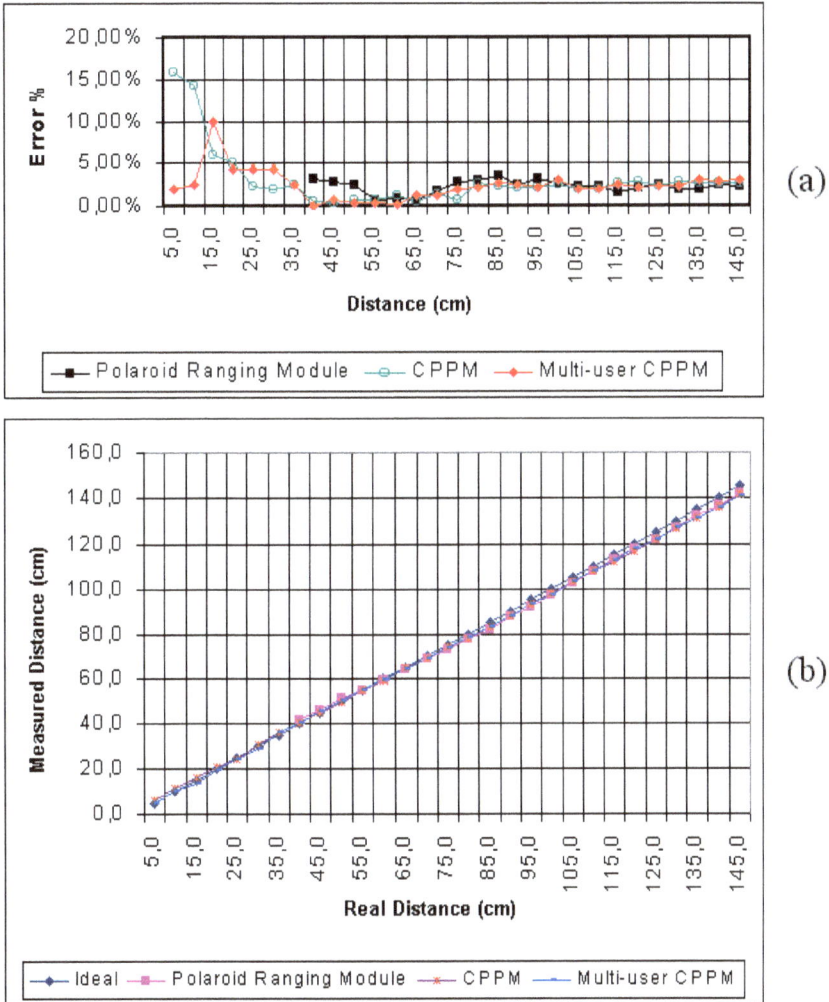

Fig. 10.11 Characterization of the sensor from 5cm to 145cm. (a) Error committed.
(b) Measurement performed.

Chapter 11

Conclusions

This book deals with Chua's circuit theory and implementation, taking into account the evolution of electronic technology, but not only this. The topics have been treated in such a way as to reflect also the new theoretical trends which have developed over the last 25 years at the borderline of Chua's circuit. As concerns the discussed implementations, experiments, performed on the Laboratory on Complex Systems of the Scuola Superiore di Catania, have been often used to give an illustration of the topics treated.

As in an evolutionary process, the main features of Chua's circuit have been maintained over time, but some of its "genes" have been changed to make it capable of showing new nonlinear phenomena, to focus on new theoretical aspects and to develop new applications. The invention of Chua's circuit can be considered a breakpoint in nonlinear electronic design. In the spirit of Thomas Kuhn's theory, Chua's circuit represents the bifurcation point that has changed the paradigm of nonlinear electronic circuit thinking.

In the original studies on chaotic systems it was maintained that the behavior of such systems was related to very rare processes and phenomena occurring in specific conditions. The term "strange attractor" could reflect this idea. The invention of Chua's circuit has proved the opposite of this; indeed a very easy explanation of chaos generation can be derived from it.

The simple components of Chua's circuit and the ease of its nonlinearity have attracted interest in chaotic circuits, encouraging discussion on chaotic behaviors and models and leading to the concept of synchronization, to the idea of chaos control and to the use of chaotic circuits in real applications. Chua's circuit can also represent a simple and fundamental tool for the investigation of new nonlinear phenomena. It is fascinating to note that a change in only a few parameters can lead to very different nonlinear phenomena. We are sure that 25 years ago Leon Chua did not foresee the richly varied behavior of his circuit and the hidden emergent phenomena contained within it.

Today, mapping real world phenomena in Chua's circuit behavior is a standard way of approaching experimental studies in research areas different from those strictly related to nonlinear electronic circuits. Chua's circuit is now a cornerstone in nonlinear electronic circuits: in the future it could be a cornerstone in understanding complex emergent phenomena.

There are a number of further appealing issues related to Chua's circuit which have not been dealt with in this book. We may think, for example, of the intriguing results obtained by using arrays of Chua's circuits, of the intrinsic capacity for generating complex dynamics of reaction diffusion equations based on it, of its use for modelling neuronal phenomena such as the birth of spiral waves and of the relationship between the bifurcation behavior of the circuit and the quantized energy level of atoms.

In the future Chua's circuit will be considered as a building block in electronic engineering studies. Now we know what a Chua's circuit is and what makes a good Chua's circuit, we are going on discovering new Chua's circuits. In writing this book, we hope that many people will have the opportunity to continue and strengthen their studies and researches on Chua's circuit.

11.1 Scheme of the book

The schematic summary of the book is set out in Table 11.1, where the main topics concerning the Chua's circuit are indicated in the columns. In particular, the following topics have been considered:

- general aspects on theory;
- Chua's circuit implementations;
- Chua's circuit implementations which can be realized in basic electronic laboratories;
- Chua's circuit implementations which can be realized in advanced laboratories;
- applications;
- state of the art and review concepts.

Table 11.1 Scheme of the book.

	Theory	Implemen-tation	Laboratory implementation	Advanced implementation	Applica-tions	Review concepts
Ch. 1	*					*
Ch. 2	*	*	*			*
Ch. 3	*	*	*			*
Ch. 4		*	*			
Ch. 5		*	*			
Ch. 6		*	*			
Ch. 7		*		*		
Ch. 8	*	*	*			
Ch. 9		*		*		
Ch. 10					*	*

Critical bibliography

Review books and special issues

Madan, R. (ed.) (1993). *Chua's Circuit: A Paradigm for Chaos* (World Scientific).

Special issue of the *Journal of Circuits, Systems and Computers*, vol. 3, no. 1, (1993): Chua's Circuit: A Paradigm for Chaos. Part 1: Introduction and Applications. Guest Editor: R. N. Madan.

Special issue of the *Journal of Circuits, Systems and Computers*, vol. 3, no. 2, (1993): Chua's Circuit: A Paradigm for Chaos. Part 2: Bifurcation and Generalizations. Guest Editor: R. N. Madan.

Proceedings of the International Symposium of Nonlinear Theory and its Applications (NOLTA '93), Honolulu, Hawai, December 5-10, 1993.

Special issue of the *IEEE Transactions on Circuits and Systems*, vol. 40, (1993): Chaos in Nonlinear Electronic Circuits. Guest Editors: L. O. Chua and M. Hasler.

Bilotta, E. and Pantano, P. (2008). *A gallery of Chua attractors* (World Scientific Series on Nonlinear Science, Series A - Vol. 61).

Theory, nonlinear phenomena, bifurcations

Altman, E. J. (1993a). Bifurcation analysis of chua's circuit with applications for low-level sensing, *Journal of Circuits and Systems Computation* **3**, pp. 63–92.

Anishchenko, V., Neiman, A. B. and Chua, L. O. (1993). Chaos-chaos intermittency and 1/f noise in chua's circuit, in R. N. Madan (ed.), *Chua's circuit: a paradigm for chaos*, World Scientific Series on Nonlinear Sciences, Series B, Vol. 1 (World Scientific, Singapore), pp. 309–324.

Anishchenko, V., Safonova, M. A. and Chua, L. O. (1992). Stochastic resonance in chua's circuit, *International Journal of Bifurcation and Chaos* **2**, pp. 397–401.

Chua, L. O. (1992). The genesis of chua's circuit, *Archiv fur Elektronik und Ubertragungstechnik* **46**, 4, pp. 250–257.

Chua, L. O. (1993). Global unfolding chua's circuit, *IEICE Transactions Fundamentals* **E76**, 5, pp. 704–734.

Chua, L. O. (1994). Chua's circuit 10 years later, *International Journal of Bifurcation and Chaos* **22**, pp. 279–305.

Chua, L. O., Komuro, M. and Matsumoto, T. (1986). The double scroll family, *IEEE Transactions on Circuits and Systems I* **33**, 11, pp. 1073–1118.

Chua, L. O. and Lin, G.-N. (1990). Canonical realization of chua's circuit family, *IEEE Transactions on Circuits and Systems* **37**, 7, pp. 885–902.

Halle, K., Chua, L., Anishchenko, V. and Safonova, M. A. (1992). Signal amplification via chaos: experimental evidence, *International Journal of Bifurcation and Chaos* **2**, pp. 1011–1020.

Kahlert, C. and Chua, L. O. (1987). Transfer maps and return maps for piecewise-linear three-region dynamical systems, *International Journal Circuit Theory and Applications* **15**, pp. 23–49.

Kennedy, M. P. (1993a). Three steps to chaos - part i: Evolution, *IEEE Transactions on Circuits and Systems I* **40**, 10, pp. 640–656.

Kennedy, M. P. (1993b). Three steps to chaos - part ii: A chua's circuit primer, *IEEE Transactions on Circuits and Systems I* **40**, 10, pp. 657–674.

Khibnik, A. I., Roose, D. and Chua, L. O. (1993a). On periodic orbits and homoclinic bifurcations in chua's circuit with a smooth nonlinearity, *International Journal of Bifurcation and Chaos* **3**, pp. 363–384.

Khibnik, A. I., Roose, D. and Chua, L. O. (1993b). On periodic orbits and homoclinic bifurcations in chua's circuit with a smooth nonlinearity, in R. N. Madan (ed.), *Chua's circuit: a paradigm for chaos*, World Scientific Series on Nonlinear Sciences, Series B, Vol. 1 (World Scientific, Singapore), pp. 145–178.

Kocarev, L., Halle, K., Eckert, K. and Chua, L. O. (1993). Experimental observations of antimonotonicity in chua's circuit, *International Journal of Bifurcations and Chaos* **3**, pp. 1051–1055.

Lukin, K. A. (1993). High frequency oscillation from chua's circuit, *Journal of Circuits and Systems Computation* **3**, pp. 627–643.

Matsumoto, T., Chua, L. O. and Komuro, M. (1985). The double scroll,

IEEE Transactions on Circuits and Systems **32**, 8, pp. 798–818.

Pivka, L. and Spany, V. (1993). Boundary surfaces and basin bifurcations in chua's circuit, *J Cir. Syst. Comput.* **3**, pp. 441–470.

Sharkovsky, A. N., Maistrenko, Y., Deregel, P. and Chua, L. O. (1993). Dry turbolence from a time-delayed chua's circuit, *Journal of Circuits and Systems Computation* **3**, pp. 645–668.

Wang, X., Zhong, G., Tang, K.-S., Man, K. and Liu, Z.-F. (2001). Generating chaos in chuas circuit via time-delay feedback, *IEEE Transactions on Circuits and Systems I* **48**, 9, pp. 1151–1156.

Yang, L. and Liao, Y. (1987). Self-similar bifurcation structures from chuas circuit, *Int. J. Circ. Th. Appl.* **15**, pp. 189–192.

Models, higher-order generalizations, implementations

Arena, P., Baglio, S., Fortuna, L. and Manganaro, G. (1996a). Generation of n-double scrolls via cellular neural networks, *International Journal of Circuit Theory and Applications* **24**, pp. 241–252.

Arena, P., Baglio, S., Fortuna, L. and Manganaro, G. (1996b). State controlled cnn: A new strategy for generating high complex dynamics, *IEICE Trans. Fundamentals* **E79-A**, pp. 1647–1657.

Awrejcewicz, J. and Calvisi, M. L. (2002). Mechanical models of chua's circuit, *International Journal of Bifurcation and Chaos* **12**, 4, pp. 671–686.

Barboza, R. (2008). Hyperchaos in a chua's circuit with two new added branches, *International Journal of Bifurcation and Chaos* **18**, 4, pp. 1151–1159.

Barboza, R. and Chua, L. O. (2008). The four-element chua's circuit, *International Journal of Bifurcation and Chaos* **18**, 4, pp. 943–955.

Cannas, B. and Cincotti, S. (2002). Hyperchaotic behaviour of two bi-directionally coupled chua's circuits, *International Journal of Circuit Theory and Applications* **30**, pp. 625–637.

Chua, L. O. (1992). The genesis of chua's circuit, *Archiv fur Elektronik und Ubertragungstechnik* **46**, 4, pp. 250–257.

Cincotti, S. and Stefano, S. D. (2004). Complex dynamical behaviours in two non-linearly coupled chua's circuits, *Chaos, Solitons and Fractals* **21**, 3, pp. 633–641.

Cruz, J. M. and Chua, L. O. (1992). A cmos ic nonlinear resistor for chua's circuit, *IEEE Transactions on Circuits and Systems I* **39**, 12, pp. 985–995.

Cruz, J. M. and Chua, L. O. (1993). An ic chip of chua's circuit, *IEEE Trans. Circuits Syst. II* **40**, 10, pp. 614–625.

Delgado-Restituto, M. and Rodriguez-Vazquez, A. (1993). A cmos monolitic chua's circuit, in R. N. Madan (ed.), *Chua's circuit: a paradigm for chaos, World Scientific Series on Nonlinear Sciences, Series B*, Vol. 1 (World Scientific, Singapore).

Fortuna, L., Frasca, M., Gioffrè, M., Rosa, M. L., Malagnino, N., Marcellino, A., Nicolosi, D., Occhipinti, L., Porro, F., Sicurella, G., Umana, E. and Vecchione, R. (2008). On the way to plastic computation, *IEEE Circuits and Systems Magazine* **8**, 3, pp. 6–18.

Fortuna, L., Frasca, M., Umana, E., Rosa, M. L., Nicolosi, D. and Sicurella, G. (2007). Organic chua's circuit, *International Journal Bifurcation and Chaos* **17**, 9, pp. 3035–3045.

Huang, A., Pivka, L., Wu, C.-W. and Franz, M. (1996). Chuas equation with cubic nonlinearity, *International Journal of Bifurcation and Chaos* **6**, pp. 2175–2222.

Kapitaniak, T., Chua, L. O. and Zhong, G.-Q. (1994). Experimental hyperchaos in coupled chua's circuits, *IEEE Transactions on Circuits and Systems I* **41**, 7, pp. 499–503.

Kennedy, M. P. (1992). Robust op amp realization of chua's circuit, *Frequenz* **46**, 3–4, pp. 66–80.

Matsumoto, T., Chua, L. O. and Tokumasu, K. (1986). Double scroll via a two-transistor circuit, *IEEE Transactions on Circuits and Systems I* **33**, 8, pp. 828–835.

Murali, K. and Lakshmanan, M. (1992). Effect of sinusoidal excitation on the chua's circuit, *IEEE Transactions on Circuits and Systems I* **39**, 4, pp. 264–270.

Murali, K. and Lakshmanan, M. (1993). Chaotic dynamics of the driven chua's circuit, *IEEE Transactions on Circuits and Systems I* **40**, 1, pp. 836–840.

Murali, K., Lakshmanan, M. and Chua, L. (1994a). Bifurcation and chaos in the simplest dissipative non-autonomous circuit, *International Journal of Bifurcation and Chaos* **4**, pp. 1511–1524.

Murali, K., Lakshmanan, M. and Chua, L. O. (1994b). The simplest dissipative non-autonomous chaotic circuit, *IEEE Transactions on Circuits and Systems I* **41**, pp. 462–463.

Ozoguz, S., Elwakil, A. S. and Salama, K. N. (2002). n-scroll chaos generator using nonlinear transconductor, *Electronics Letters* **38**, pp. 685–686.

Rodriguez-Vazquez, A. and Delgado-Restituto, M. (1993). Cmos design of chaotic oscillators using state variables: A monolithic chua's circuit, *IEEE Transactions on Circuits and Systems II* **40**, 10, pp. 596–613.

Shi, Z. and Ran, L. (2004). Tunnel diode based chua's circuit, in *IEEE 6th CAS Symp. on Emerging Technologies: Mobile and Wireless Comm., Shangai, China, May 31-June2*, pp. 217–220.

Suykens, J. A. K. and Chua, L. O. (1997). n-double scroll hypercubes in 1-d cnns, *International Journal of Bifurcation and Chaos* **7**, 8, pp. 1873–1885.

Suykens, J. A. K., Huang, A. and Chua, L. O. (1997). A family of n-scroll attractors from a generalized chuas circuit, *Archiv fur Elektronik und Ubertragungstechnik* **51**, 3, pp. 131–138.

Suykens, J. K. and Vandewalle, J. (1993). Generation of n-double scrolls (n=1;2;3;4; ...), *IEEE Transactions on Circuits and Systems I* **40**, pp. 861–867.

Torres, L. A. B. and Aguirre, L. A. (2000). Inductorless chua's circuit, *Electronics Letters* **36**, 23, pp. 1915–1916.

Yalcin, M. E., Suykens, J. A. K. and Vandewalle, J. (2000). Experimental confirmation of 3- and 5-scroll attractors from a generalized chua's circuit, **47**, 3, pp. 425–429.

Yalcin, M. E., Suykens, J. A. K., Vandewalle, J. and Ozoguz, S. (2002). Families of scroll grid attractors, *International Journal of Bifurcation and Chaos* **12**, pp. 23–41.

Zhong, G. Q. (1994). Implementation of chua's circuit with a cubic nonlinearity, *IEEE Transactions on Circuits and Systems I* **41**, 12, pp. 934–941.

Applications

Buscarino, A., Fortuna, L., Frasca, M. and Muscato, G. (2007b). Chaos does help motion control, *International Journal of Bifurcation and Chaos* **17**, 10, pp. 3577–3581.

Dedieu, H., Kennedy, M. P. and Hasler, M. (1993). Chaos shift keying: Modulation and demodulation of a chaotic carrier using self-synchronizing chua's circuits, *IEEE Transactions on Circuits and Systems II* **40**, 10, pp. 634–642.

Fortuna, L., Frasca, M. and Rizzo, A. (2001b). Chaos preservation through continuous chaotic pulse position modulation, in *IEEE Interna-*

tional Symposium on Circuits and Systems, ISCAS'01, Sidney, Australia.

Fortuna, L., Frasca, M. and Rizzo, A. (2003a). Chaotic pulse position modulation to improve the efficiency of sonar sensors, *IEEE Transactions on Instrumentation and Measurement* **52**, 6, pp. 1809–1814.

Kocarev, L., Halle, K. S., Eckert, K., Chua, L. . and Parlitz, U. (1992). Experimental demonstration of secure communications via chaotic synchronization, *Int. J. Bifurcation and Chaos* **2**, 3, pp. 709–713.

Maggio, G. M., Rulkov, N., Sushchik, M., Tsimring, L., Volkovskii, A., Abarbanel, H., Larson, L. and Yao, K. (1999). Chaotic pulse-position modulation for ultrawide-band communication systems, in *Proc. UWB'99, Washington D.C., Sept 28-30.*

Mayer-Kress, G., Choi, I., Weber, N., Bargar, R. and Hubler, A. (1993). Musical signals from chua's circuit, *IEEE Transactions on Circuits and Systems II* **40**, pp. 688–695.

Pinkney, J. Q., Camwell, P. L. and Davies, R. (1995). Chaos shift keying communications using self-synchronizing chua oscillators, *Electronic Letters* **31**, 13, pp. 1021–1022.

Yang, T. and Chua, L. O. (1996). Secure communication via chaotic parameter modulation, *IEEE Transactions on Circuits and Systems I* **43**, 9, pp. 817–819.

Yang, T., Wu, C. W. and Chua, L. O. (1997). Cryptography based on chaotic systems, *IEEE Transactions on circuits and systems I* **44**, 5, pp. 469–472.

Arrays, networks

Arena, P., Caponetto, R., Fortuna, L. and Rizzo, A. (2000). Nonorganized deterministic dissymmetries induce regularity in spatiotemporal dynamics, *International Journal of Bifucation and Chaos* **10**, 1, pp. 73–85.

Belykh, V. N., Verichev, N. N., Kocarev, L. and Eckert, K. (1993). On chaotic synchronization in a linear arrays of chua's circuits, in R. N. Madan (ed.), *Chua's circuit: a paradigm for chaos, World Scientific Series on Nonlinear Sciences, Series B*, Vol. 1 (World Scientific, Singapore), pp. 325–335.

Dabrowski, A. M., Dabroski, W. R. and Ogorzalek, M. J. (1993). Dynamic phenomena in chain interconnections of chua's circuits, *IEEE Trans. Circuits Syst. I* **40**, 11, pp. 868–871.

De Castro, M., Perez-Munuzuri, V., Chua, L. O. and Perez-Villar, V.

(1995). Complex feigenbaum's scenario in discretely-coupled arrays of non-linear circuits, *Int. J. Bifurcation and Chaos* **5**, 3, pp. 859–868.

Munuzuri, A. P., Perez-Munuzuri, V., Gomez-Gesteira, L. O., M.and Chua and Perez-Villar, V. (17–50). Spatiotemporal structures in discretely-coupled arrays of nonlinear circuits: a review, *Int. J. Bifurcation and Chaos* **5**, 1, p. 1995.

Perez-Munuzuri, A., Perez-Munuzuri, V., Gomez-Gesteria, M., Chua, L. O. and Perez-Villar, V. (1995). Spatiotemporal structures in discretely-coupled arrays of nonlinear circuits: a review, *International Journal of Bifurcation and Chaos* **5**, 1, pp. 17–50.

Perez-Munuzuri, A., Perez-Munuzuri, V. and Perez-Villar, V. (1993a). Spiral waves on a two-dimensional array of nonlinear circuits, *IEEE Transactions on Circuits and Systems I* **40**, pp. 872–877.

Perez-Munuzuri, A., Perez-Villar, V. and Chua, L. O. (1993b). Autowaves for image processing on a two-dimensional cnn array of excitable nonlinear circuits: flat and wrinkled labyrinths, *IEEE Transactions on Circuits and Systems I* **40**, pp. 174–193.

Perez-Munuzuri, V., Gomez-Gesteira, V., Perez-Villar, V. and Chua, L. O. (1993c). Traveling wave propagation in a one-dimensional fluctuating medium, *International Journal of Bifurcation and Chaos* **3**, pp. 211–215.

Perez-Munuzuri, V., Perez-Villar, V. and Chua, L. O. (1992). Propagation failure in linear arrays of chua's circuits, *Int. J. Bifurcation and Chaos* **2**, 2, pp. 403–406.

Perez-Munuzuri, V., Perez-Villar, V. and Chua, L. O. (1993d). Travelling wave front and its failure in a one-dimensional array of chua's circuits, *Journal of Circuits and Systems Computation* **3**, pp. 215–229.

Sanchez, E., Matias, M. A. and Perez-Munuzuri, V. (2000). Chaotic synchronization in small assemblies of driven chua's circuits, *IEEE Transactions on Circuits and Systems I* **47**, 5, pp. 644–654.

Synchronization

Arena, P., Buscarino, A., Fortuna, L. and Frasca, M. (2006). Separation and synchronization of pwl chaotic systems, *Physical Review E* **74**, pp. 026212–1–7.

Belykh, V. N., Verichev, N. N., Kocarev, L. and Eckert, K. (1993). On chaotic synchronization in a linear arrays of chua's circuits, in R. N. Madan (ed.), *Chua's circuit: a paradigm for chaos*, *World Scientific Series* on *Nonlinear Sciences, Series B*, Vol. 1 (World Scientific, Singapore), pp. 325–335.

Buscarino, A., Fortuna, L. and Frasca, M. (2007a). Separation and synchronization of chaotic signals by optimization, *Physical Review E* **75**, pp. 016215–1–9.

Buscarino, A., Fortuna, L. and Frasca, M. (2008). Experimental separation of chaotic signals through synchronization, *Philosophical Transactions of the Royal Society A* **386**, 1865, pp. 569–577.

Fortuna, L., Frasca, M. and Rizzo, A. (2003b). Experimental pulse synchronisation of two chaotic circuits, *Chaos, Solitons and Fractals* **17**, pp. 355–361.

Li, C. and Liao, X. (2004). Lag synchronization of rossler system and chua circuit via a scalar signal, *Physics Letters A* **329**, 4-5, pp. 301–308.

Sanchez, E., Matias, M. A. and Perez-Munuzuri, V. (2000). Chaotic synchronization in small assemblies of driven chua's circuits, *IEEE Transactions on Circuits and Systems I* **47**, 5, pp. 644–654.

Tang, K. S., Zhong, G. Q., Chen, G. and Man, K. F. (2001). Generation of n-scroll attractors via sine function, *IEEE Transactions on Circuits and Systems I* **48**, pp. 1369–1372.

Wang, S.-H., Liu, W.-Q., Ma, B.-J., Xiao, J.-H. and Jiang, D. Y. (2005). Phase synchronization of chua circuit induced by the periodic signals, *Chinese Physics* **14**, 1, pp. 55–60.

Yang, T. and Chua, L. O. (1997). Impulsive stabilization for control and synchronization of chaotic systems: theory and applications to secure communication, *IEEE Transactions on circuits and systems I* **44**, 10, pp. 976–988.

Control

Arena, P., Fortuna, L. and Frasca, M. (2002). Chaos control by using motor maps, *Chaos* **12**, 3, pp. 559–573.

Celka, P. (1994). Experimental verification of pyragas's chaos control method applied to chua's circuit, *International Journal of Bifurcation and Chaos* **4**, 6, pp. 1703–1706.

Chang, Y.-C. (2001). A robust tracking control for chaotic chua's circuits via fuzzy approach, *IEEE Transactions on circuits and systems I* **48**, 7, pp. 889–895.

Chen, G. (1993). Controlling chua's global unfolding circuit family,

IEEE Transactions on Circuits and Systems I **40**, 11, pp. 829–832.

Chen, G. and Dong, X. (1993a). Controlling chua's circuit, *Journal of Circuits and Systems Computation* **3**, 1, pp. 139–149.

Chen, G. and Dong, X. (1993b). Controlling chua's circuit, in R. N. Madan (ed.), *Chua's circuit: a paradigm for chaos, World Scientific Series on Nonlinear Sciences, Series B*, Vol. 1 (World Scientific, Singapore), pp. 481–491.

Chen, G. and Dong, X. (1993c). From chaos to order - perspective and methologies in controlling chaotic nonlinear dynamical systems, *International Journal of Bifurcation and Chaos* **3**, 6, pp. 1363–1409.

Chen, G. and Dong, X. (1998). *From Chaos To Order: Methodologies, Perspectives and Applications* (World Scientific, Singapore).

Ge, S. S. and Wang, C. (2000). Adaptive control of uncertain chua's circuits, *IEEE Transactions on circuits and systems I* **47**, 9, pp. 1397–1402.

Genesio, R. and Tesi, A. (1993). Distorsion control of chaotic systems: the chua's circuit, in R. N. Madan (ed.), *Chua's circuit: a paradigm for chaos, World Scientific Series on Nonlinear Sciences, Series B*, Vol. 1 (World Scientific, Singapore), pp. 514–534.

Hartley, T. and Mossayebi, F. (1993a). Control of chua's circuit, *Journal of Circuits and Systems Computation* **3**, pp. 173–194.

Hartley, T. and Mossayebi, F. (1993b). Control of chua's circuit, in R. N. Madan (ed.), *Chua's circuit: a paradigm for chaos, World Scientific Series on Nonlinear Sciences, Series B*, Vol. 1 (World Scientific, Singapore), pp. 492–513.

Hwang, C. C. (1997). A linear continuous feedback control of chua's circuit, *Chaos Solitons and Fractals* **8**, 9, pp. 1507–1515.

Hwang, C. C., Chow, H. and Wang, Y. (1996). A new feedback control of a modified chuas circuit system, *Physica D* **92**, pp. 95–100.

Johnson, G., Tigner, T. E. and Hunt, E. R. (1993). Controlling chaos in chua's circuit, in R. N. Madan (ed.), *Chua's circuit: a paradigm for chaos, World Scientific Series on Nonlinear Sciences, Series B*, Vol. 1 (World Scientific, Singapore), pp. 449–457.

Johnston, G. and Hunt, E. R. (1993). Derivative control of the steady state in chua's circuit driven in the chaotic region, *IEEE Transactions on circuits and systems I* **40**, 11, pp. 833–835.

Kapitaniak, T., Kocarev, L. and Chua, L. . (1993). Controlling chaos without feedback and control signals, *International Journal of Bifurcation and Chaos* **3**, 2, pp. 459–468.

Li, Z. G., Wen, C. Y., Soh, Y. C. and Xie, W. X. (2001). The stabilization and synchronization of chua's oscillators via impulsive control, *IEEE Transactions on circuits and systems I* **48**, 11, pp. 1351–1355.

Ogorzalek, M. J. (1993b). Taming chaos: Part ii control, *IEEE Transactions on Circuits and Systems I* **40**, 10, pp. 700–706.

Puebla, H., Alvarez-Ramirez, J. and Cervantes, I. (2003). A simple tracking control for chua's circuit, *IEEE Transactions on circuits and systems I* **50**, 2, pp. 280–284.

Torres, L. and Aguirre, L. (1999). Extended chaos control method applied to chua circuit, *Electronics Letters* **35**, 10, pp. 768–770.

Yang, T. and Chua, L. O. (1997). Impulsive stabilization for control and synchronization of chaotic systems: theory and applications to secure communication, *IEEE Transactions on circuits and systems I* **44**, 10, pp. 976–988.

Zhang, T. and Feng, G. (2007). Output tracking of piecewise-linear systems via error feedback regulator with application to synchronization of nonlinear chua's circuit, *IEEE Transactions on Circuits and Systems I* **54**, 8, pp. 1852–1863.

Bibliography

Altman, E. J. (1993a). Bifurcation analysis of chua's circuit with applications for low-level sensing, *Journal of Circuits and Systems Computation* **3**, pp. 63–92.

Altman, E. J. (1993b). Normal form analysis of chua's circuit with applications for trajectory recognition, *IEEE Transactions on Circuits and Systems II* **40**, pp. 675–682.

Anadigm (2003). URL www.anadigm.com.

Andò, B., Baglio, S., Graziani, S. and Pitrone, N. (2000). Cnns for noise generation in dithered transducers, in *Proceedings of the 17th IEEE Instrumentation and Measurement Technology Conference, 2000 (IMTC 2000)*, Vol. 2, pp. 1071–1076.

Andò, B. and Graziani, S. (2000). *Stochastic resonance - Theory and applications* (Kluwer Academic Publishers, Boston).

Anishchenko, V., Astakhov, V., Neiman, A., Vadivasova, T. and Schimansky-Geier, L. (2003). *Nonlinear Dynamics of Chaotic and Stochastic Systems* (Springer Series in Synergetics, Boston).

Anishchenko, V., Neiman, A. B. and Chua, L. O. (1993). Chaos-chaos intermittency and 1/f noise in chua's circuit, in R. N. Madan (ed.), *Chua's circuit: a paradigm for chaos, World Scientific Series on Nonlinear Sciences, Series B*, Vol. 1 (World Scientific, Singapore), pp. 309–324.

Anishchenko, V., Safonova, M. A. and Chua, L. O. (1992). Stochastic resonance in chua's circuit, *International Journal of Bifurcation and Chaos* **2**, pp. 397–401.

Arena, P., Baglio, S., Fortuna, L. and Manganaro, G. (1995a). Chua's circuit can be generated by cnn cells, *IEEE Transactions on Circuits and Systems-I* **42**, 2, pp. 123–125.

Arena, P., Baglio, S., Fortuna, L. and Manganaro, G. (1995b). Simplified scheme for realisation of chua oscillator by using sc-cnn cells, *Electronics Letters* **31**, 21, pp. 1794–1795.

Arena, P., Baglio, S., Fortuna, L. and Manganaro, G. (1996a). Generation of n-double scrolls via cellular neural networks, *International Journal of Circuit Theory and Applications* **24**, pp. 241–252.

193

Arena, P., Baglio, S., Fortuna, L. and Manganaro, G. (1996b). State controlled cnn: A new strategy for generating high complex dynamics, *IEICE Trans. Fundamentals* **E79-A**, pp. 1647–1657.

Arena, P., Bucolo, M., Fazzino, S., Fortuna, L. and Frasca, M. (2005). The cnn paradigm: Shapes and complexity, *International Journal of Bifurcation and Chaos* **15**, 7, pp. 2063–2090.

Arena, P., Buscarino, A., Fortuna, L. and Frasca, M. (2006). Separation and synchronization of pwl chaotic systems, *Physical Review E* **74**, pp. 026212-1–7.

Arena, P., Caponetto, R., Fortuna, L. and Rizzo, A. (2000). Nonorganized deterministic dissymmetries induce regularity in spatiotemporal dynamics, *International Journal of Bifucation and Chaos* **10**, 1, pp. 73–85.

Arena, P., Castorina, S., Fortuna, L., Frasca, M. and Rizzo, A. (2004). An integrated chua's cell for the implementation of a chua's array, *International Journal Bifurcation and Chaos* **14**, 1, pp. 93–106.

Arena, P. and Fortuna, L. (2000). Collective behaviour in cellular neural networks to model the central pattern generator, *International Journal of Systems Science* **31**, 7, pp. 827–841.

Arena, P., Fortuna, L. and Frasca, M. (2002). Chaos control by using motor maps, *Chaos* **12**, 3, pp. 559–573.

Arneodo, A., Coulett, P. and Tresser, C. (1982). Oscillators with chaotic behavior: An illustration of a theorem by shilnikov, *J. Stat. Phys.* **27**, pp. 171–182.

Awrejcewicz, J. and Calvisi, M. L. (2002). Mechanical models of chua's circuit, *International Journal of Bifurcation and Chaos* **12**, 4, pp. 671–686.

Baird, B., Hirsch, M. and Eeckman, F. (1993). A neural network associative memory for handwritten character recognition using multiple chua attractor, *IEEE Transactions on Circuits and Systems II* **40**, pp. 667–674.

Baptista, M. S., Silva, T. P., Sartorelli, J. C., Caldas, I. L. and Rosa, E. J. (2003). Phase synchronization in the perturbed chua circuit, *Physical Review E* **67**, 5, pp. 56212-1–5.

Barboza, R. (2008). Hyperchaos in a chua's circuit with two new added branches, *International Journal of Bifurcation and Chaos* **18**, 4, pp. 1151–1159.

Barboza, R. and Chua, L. O. (2008). The four-element chua's circuit, *International Journal of Bifurcation and Chaos* **18**, 4, pp. 943–955.

Barone, K. and Singh, S. N. (2002). Adaptive feedback linearizing control of chua's circuit, *International Journal of Bifurcation and Chaos* **12**, 7, pp. 1599–1604.

Belykh, V. N., Verichev, N. N., Kocarev, L. and Eckert, K. (1993). On chaotic synchronization in a linear arrays of chua's circuits, in R. N. Madan (ed.), *Chua's circuit: a paradigm for chaos, World Scientific Series on Nonlinear Sciences, Series B*, Vol. 1 (World Scientific, Singapore), pp. 325–335.

Benzi, R., Sutera, A. and Vulpiani, A. (1981). The mechanism of stochastic resonance, *J. Phys. A: Math. Gen.* **14**, pp. 453–457.

Bilotta, E., Gervasi, S. and Pantano, P. (2005). Reading complexity in chua's oscillator through music. part i: A new way of understanding chaos, *International Journal of Bifurcation and Chaos* **15**, 2, pp. 253–382.

Bilotta, E. and Pantano, P. (2008). *A gallery of Chua attractors* (World Scientific Series on Nonlinear Science, Series A - Vol. 61).

Birk, C. (1998). Evaluation of filters implemented using a field programmable analog array, in *WMC'98-ICSEE, Jan. 11-14, San Diego*.

Boccaletti, S., Grebogi, C., Lai, Y.-C., Mancini, H. and Maza, D. (2000). The control of chaos: theory and applications, *Physics report* **329**, pp. 103–197.

Boccaletti, S., Kurths, J., Osipov, G., Valladares, D. L. and Zhou, C. S. (2002). The synchronization of chaotic systems, *Physics Reports* **366**, pp. 1–101.

Boccaletti, S., Latora, V., Moreno, Y., Chavez, M. and Hwang, D.-U. (2006). Complex networks: Structure and dynamics, *Physics Reports* **424**, pp. 175–308.

Borenstein, J. and Koren, Y. (1994). Error eliminating rapid ultrasonic firing for mobile robot obstacle avoidance, *EEE Trans. on Robotics and Automation* **11**, 1, pp. 132–138.

Brown, R., Leeuw, D. M., Havinga, E. E. and Pomp, A. (1994). A universal relation between conductivity and field-effect mobility in doped amorphous organic semiconductors, *Synth. Mat* **68**, pp. 65–70.

Buscarino, A., Fortuna, L. and Frasca, M. (2007a). Separation and synchronization of chaotic signals by optimization, *Physical Review E* **75**, pp. 016215-1–9.

Buscarino, A., Fortuna, L. and Frasca, M. (2008). Experimental separation of chaotic signals through synchronization, *Philosophical Transactions of the Royal Society A* **386**, 1865, pp. 569–577.

Buscarino, A., Fortuna, L., Frasca, M. and Muscato, G. (2007b). Chaos does help motion control, *International Journal of Bifurcation and Chaos* **17**, 10, pp. 3577–3581.

Cannas, B. and Cincotti, S. (2002). Hyperchaotic behaviour of two bi-directionally coupled chua's circuits, *International Journal of Circuit Theory and Applications* **30**, pp. 625–637.

Cantelli, L., Fortuna, L., Frasca, M. and Rizzo, A. (2001). Frequency switched chua's circuit: Experimental dynamics characterization, *International Journal Bifurcation and Chaos* **11**, 1, pp. 231–239.

Caponetto, R., Mauro, A. D., Fortuna, L. and Frasca, M. (2005). Field programmable analog arrays to implement programmable chua's circuit, *International Journal Bifurcation and Chaos* **15**, 5, pp. 1829–1836.

Celka, P. (1994). Experimental verification of pyragas's chaos control method applied to chua's circuit, *International Journal of Bifurcation and Chaos* **4**, 6, pp. 1703–1706.

Chang, Y.-C. (2001). A robust tracking control for chaotic chua's circuits via fuzzy approach, *IEEE Transactions on circuits and systems I* **48**, 7, pp. 889–895.

Chen, G. (1993). Controlling chua's global unfolding circuit family, *IEEE Transactions on Circuits and Systems I* **40**, 11, pp. 829–832.

Chen, G. and Dong, X. (1993a). Controlling chua's circuit, *Journal of Circuits and Systems Computation* **3**, 1, pp. 139–149.

Chen, G. and Dong, X. (1993b). Controlling chua's circuit, in R. N. Madan (ed.),

Chua's circuit: a paradigm for chaos, World Scientific Series on Nonlinear Sciences, Series B, Vol. 1 (World Scientific, Singapore), pp. 481–491.

Chen, G. and Dong, X. (1993c). From chaos to order - perspective and methologies in controlling chaotic nonlinear dynamical systems, *International Journal of Bifurcation and Chaos* **3**, 6, pp. 1363–1409.

Chen, G. and Dong, X. (1998). *From Chaos To Order: Methodologies, Perspectives and Applications* (World Scientific, Singapore).

Chua, L. O. (1971). Memristor - the missing circuit element, *IEEE Transactions on Circuit Theory* **18**, 5, pp. 507–519.

Chua, L. O. (1992). The genesis of chua's circuit, *Archiv fur Elektronik und Ubertragungstechnik* **46**, 4, pp. 250–257.

Chua, L. O. (1993). Global unfolding chua's circuit, *IEICE Transactions Fundamentals* **E76**, 5, pp. 704–734.

Chua, L. O. (1994). Chua's circuit 10 years later, *International Journal of Bifurcation and Chaos* **22**, pp. 279–305.

Chua, L. O. (1998). *CNN: A Paradigm for Complexity* (World Scientific, Singapore/River Edge: NJ).

Chua, L. O., Itoh, M., Kocarev, L. and Eckert, K. (1993a). Chaos synchronization in chua's circuit, in R. N. Madan (ed.), *Chua's circuit: a paradigm for chaos, World Scientific Series on Nonlinear Sciences, Series B*, Vol. 1 (World Scientific, Singapore), pp. 309–324.

Chua, L. O., Itoh, M., Kocarev, L. and Eckert, K. (1993b). Chaos synchronization in chua's circuit, *Journal of Circuits, Systems, and Computers* **3**, 1, pp. 93–108.

Chua, L. O., Kocarev, L., Eckert, K. and Itoh, M. (1992). Experimental chaos synchronization in chua's circuit, *International Journal of Bifurcation and Chaos* **2**, 3, pp. 705–708.

Chua, L. O., Komuro, M. and Matsumoto, T. (1986). The double scroll family, *IEEE Transactions on Circuits and Systems I* **33**, 11, pp. 1073–1118.

Chua, L. O. and Lin, G.-N. (1990). Canonical realization of chua's circuit family, *IEEE Transactions on Circuits and Systems* **37**, 7, pp. 885–902.

Chua, L. O. and Roska, T. (1993). The CNN paradigm, *IEEE Trans. Circuits and Systems I* **40**, pp. 147–156.

Chua, L. O. and Roska, T. (2005). *Cellular Neural Networks and Visual Computing: Foundations and Applications* (Cambridge University Press).

Chua, L. O. and Yang, L. (1988a). Cellular neural networks: applications, *IEEE Trans. Circuits and Systems I* **35**, pp. 1247–1290.

Chua, L. O. and Yang, L. (1988b). Cellular neural networks: theory, *IEEE Trans. Circuits and Systems I* **35**, pp. 1257–1272.

Chua, L. O., Yang, T., Zhong, G.-Q. and Wu, C.-W. (1996). Synchronization of chua's circuits with time-varying channels and parameters, *IEEE Transactions on Circuits and Systems I* **43**, 10, pp. 862–868.

Cincotti, S. and Stefano, S. D. (2004). Complex dynamical behaviours in two non-linearly coupled chua's circuits, *Chaos, Solitons and Fractals* **21**, 3, pp. 633–641.

Conrad, J. M. and Mills, J. (1997). *Stiquito: advanced experiment with a simple and inexpensive robot* (IEEE World Computer Society, Princeton).

Cruz, J. M. and Chua, L. O. (1992). A cmos ic nonlinear resistor for chua's circuit, *IEEE Transactions on Circuits and Systems I* **39**, 12, pp. 985–995.

Cruz, J. M. and Chua, L. O. (1993). An ic chip of chua's circuit, *IEEE Trans. Circuits Syst. II* **40**, 10, pp. 614–625.

Dabrowski, A. M., Dabroski, W. R. and Ogorzalek, M. J. (1993). Dynamic phenomena in chain interconnections of chua's circuits, *IEEE Trans. Circuits Syst. I* **40**, 11, pp. 868–871.

Dana, S. K. and Roy, P. K. (2003). Experimental evidence on intermittent lag synchronization in coupled chua's oscillators, *American Institute of Physics Conference Proceedings* **676**, p. 363.

Dana, S. K., Roy, P. K., Mukhopadhyay, B. and Chakraborty, S. (2003). Phase synchronization of shil'nikov chaos in coupled chua's oscillators, *American Institute of Physics Conference Proceedings* **676**, pp. 62–67.

Dario, P., Guglielmelli, E. and Allotta, B. (1994). Robotics in medicine, in *Proceedings of the IEEE/RSJ/GI International Conference on Intelligent Robots and Systems '94*, Vol. 2, pp. 739–752.

Dawson, P., Grebogi, C., Yorke, J., Kan, I. and Kocak, H. (1992). Antimonicity - inevitable reversal of period-doubling cascades, *Physical Letters A* **162**, pp. 249–252.

De Ambroggi, F., Fortuna, L. and Muscato, G. (1997). Plif: Piezo light intelligent flea. new micro-robots controlled by self-learning techniques, in *Proc. of IEEE Int. Conf. Robotics and Automation.*

De Castro, M., Perez-Munuzuri, V., Chua, L. O. and Perez-Villar, V. (1995). Complex feigenbaum's scenario in discretely-coupled arrays of nonlinear circuits, *Int. J. Bifurcation and Chaos* **5**, 3, pp. 859–868.

Dedieu, H., Kennedy, M. P. and Hasler, M. (1993). Chaos shift keying: Modulation and demodulation of a chaotic carrier using self-synchronizing chua's circuits, *IEEE Transactions on Circuits and Systems II* **40**, 10, pp. 634–642.

Delgado-Restituto, M. and Rodriguez-Vazquez, A. (1993). A cmos monolitic chua's circuit, in R. N. Madan (ed.), *Chua's circuit: a paradigm for chaos, World Scientific Series on Nonlinear Sciences, Series B*, Vol. 1 (World Scientific, Singapore).

der Pol, B. V. and der Mark, J. V. (1927). Frequency demultiplication, *Nature* **120**, pp. 363–364.

Dimitrakopoulos, C. and Mascaro, D. J. (2001). Organic thin-film transistors: a review of recent advances, *IBM J. Res. & Dev.* **45**, pp. 11–27.

D'Mello, D. R. and Gulak, P. G. (1998). Design approaches to field-programmable analog integrated circuits, *Analog Integrated Circuits and Signal Processing* **17**, pp. 7–34.

Duffing, G. (1918). *Erzwungene Schwingungen bei Veränderlicher Eigenfrequenz* (F. Vieweg u. Sohn, Braunschweig DE).

Ebisawa, F., Kurokawa, T. and Nara, S. (1983). Electrical properties of polyacetylene/polysiloxane interface, *Jour. of Appl. Phys.* **54**, 6, p. 3255.

Everett, H. R. (1995). *Sensors for Mobile Robots - Theory and Application* (A.

K. Peters Ltd., Natick, MA).

Fortuna, L., Arena, P., Balya, D. and Zarandy, A. (2001a). Cellular neural networks: a paradigm for nonlinear spatio-temporal processing, *IEEE Circuits and Systems Magazine* **1**, 4, pp. 6–21.

Fortuna, L., Frasca, M., Gioffrè, M., Rosa, M. L., Malagnino, N., Marcellino, A., Nicolosi, D., Occhipinti, L., Porro, F., Sicurella, G., Umana, E. and Vecchione, R. (2008). On the way to plastic computation, *IEEE Circuits and Systems Magazine* **8**, 3, pp. 6–18.

Fortuna, L., Frasca, M. and Rizzo, A. (2001b). Chaos preservation through continuous chaotic pulse position modulation, in *IEEE International Symposium on Circuits and Systems, ISCAS'01, Sidney, Australia*.

Fortuna, L., Frasca, M. and Rizzo, A. (2003a). Chaotic pulse position modulation to improve the efficiency of sonar sensors, *IEEE Transactions on Instrumentation and Measurement* **52**, 6, pp. 1809–1814.

Fortuna, L., Frasca, M. and Rizzo, A. (2003b). Experimental pulse synchronisation of two chaotic circuits, *Chaos, Solitons and Fractals* **17**, pp. 355–361.

Fortuna, L., Frasca, M., Umana, E., Rosa, M. L., Nicolosi, D. and Sicurella, G. (2007). Organic chua's circuit, *International Journal Bifurcation and Chaos* **17**, 9, pp. 3035–3045.

Gammaitoni, L., Hänggi, P., Jung, P. and Marchesoni, F. (1998). Stochastic resonance, *Reviews of Modern Physics* **70**, 1, pp. 223–287.

Gamota, D. R., Brazis, P., Kalyanasundaram, K. and Zhang, J. (eds.) (2004). *Printed Organic and Molecular Electronics* (Kluwer academic Publishers).

Ge, S. S. and Wang, C. (2000). Adaptive control of uncertain chua's circuits, *IEEE Transactions on circuits and systems I* **47**, 9, pp. 1397–1402.

Genesio, R. and Tesi, A. (1993). Distorsion control of chaotic systems: the chua's circuit, in R. N. Madan (ed.), *Chua's circuit: a paradigm for chaos, World Scientific Series on Nonlinear Sciences, Series B*, Vol. 1 (World Scientific, Singapore), pp. 514–534.

Gomes, I., Mirasso, C. R., Toral, R. and Calvo, O. (2003). Experimental study of high frequency stochastic resonance in chua circuits, *Physica A* **327**, 1-2, pp. 115–119.

Grebogi, C., Ott, E., Pelikan, S. and Yorke, J. A. (1984). Strange attractors that are not chaotic, *Physica D* **13**, 1-2, pp. 261–268.

Gregorian, G. C., R.and Temes (1986). *Analog MOS Integrated Circuit For Signal Processing* (Wiley-Interscience).

Guckenheimer, J. and Holmes, P. (1983). *Nonlinear Oscillations, Dynamical Systems, and Bifurcations of Vector Fields* (Springer Verlag, New York).

Hagen, K., Gundlach, D. J. and Jackson, T. N. (1999). Fast organic thin-film transistor circuits, *IEEE Electron Devices Letters* **20**, pp. 289–291.

Halle, K., Chua, L., Anishchenko, V. and Safonova, M. A. (1992). Signal amplification via chaos: experimental evidence, *International Journal of Bifurcation and Chaos* **2**, pp. 1011–1020.

Hartley, T. and Mossayebi, F. (1993a). Control of chua's circuit, *Journal of Circuits and Systems Computation* **3**, pp. 173–194.

Hartley, T. and Mossayebi, F. (1993b). Control of chua's circuit, in R. N. Madan (ed.), *Chua's circuit: a paradigm for chaos*, World Scientific Series on Nonlinear Sciences, Series B, Vol. 1 (World Scientific, Singapore), pp. 492–513.

Hu, G., Pivka, L. and Zheleznyak, A. L. (1995). Synchronization of a one-dimensional array of chua's circuits by feedback control and noise, *IEEE Transactions on Circuits and Systems I* **42**, 10, pp. 736–740.

Huang, A., Pivka, L., Wu, C.-W. and Franz, M. (1996). Chuas equation with cubic nonlinearity, *International Journal of Bifurcation and Chaos* **6**, pp. 2175–2222.

Hulub, M., Frasca, M., Fortuna, L. and Arena, P. (2006). Implementation and synchronization of a 3x3 grid scroll attractor with analog programmable devices, *Chaos* **16**, 1, pp. 013121–1–5.

Hwang, C. C. (1997). A linear continuous feedback control of chua's circuit, *Chaos Solitons and Fractals* **8**, 9, pp. 1507–1515.

Hwang, C. C., Chow, H. and Wang, Y. (1996). A new feedback control of a modified chuas circuit system, *Physica D* **92**, pp. 95–100.

Johnson, G., Tigner, T. E. and Hunt, E. R. (1993). Controlling chaos in chua's circuit, in R. N. Madan (ed.), *Chua's circuit: a paradigm for chaos*, World Scientific Series on Nonlinear Sciences, Series B, Vol. 1 (World Scientific, Singapore), pp. 449–457.

Johnston, G. and Hunt, E. R. (1993). Derivative control of the steady state in chua's circuit driven in the chaotic region, *IEEE Transactions on circuits and systems I* **40**, 11, pp. 833–835.

Jörg, K. W. and Berg, M. (1998). Mobile robot sensing with pseudo-random codes, in *1998 IEEE International Conference on Robotics and Automation*, pp. 2807–2812.

Kahlert, C. and Chua, L. O. (1987). Transfer maps and return maps for piecewise-linear three-region dynamical systems, *International Journal Circuit Theory and Applications* **15**, pp. 23–49.

Kapitaniak, T., Chua, L. O. and Zhong, G.-Q. (1994). Experimental hyperchaos in coupled chua's circuits, *IEEE Transactions on Circuits and Systems I* **41**, 7, pp. 499–503.

Kapitaniak, T., Kocarev, L. and Chua, L. . (1993). Controlling chaos without feedback and control signals, *International Journal of Bifurcation and Chaos* **3**, 2, pp. 459–468.

Kennedy, M. P. (1992). Robust op amp realization of chua's circuit, *Frequenz* **46**, 3–4, pp. 66–80.

Kennedy, M. P. (1993a). Three steps to chaos - part i: Evolution, *IEEE Transactions on Circuits and Systems I* **40**, 10, pp. 640–656.

Kennedy, M. P. (1993b). Three steps to chaos - part ii: A chua's circuit primer, *IEEE Transactions on Circuits and Systems I* **40**, 10, pp. 657–674.

Khibnik, A. I., Roose, D. and Chua, L. O. (1993a). On periodic orbits and homoclinic bifurcations in chua's circuit with a smooth nonlinearity, *International Journal of Bifurcation and Chaos* **3**, pp. 363–384.

Khibnik, A. I., Roose, D. and Chua, L. O. (1993b). On periodic orbits and homoclinic bifurcations in chua's circuit with a smooth nonlinearity, in R. N.

Madan (ed.), *Chua's circuit: a paradigm for chaos, World Scientific Series on Nonlinear Sciences, Series B*, Vol. 1 (World Scientific, Singapore), pp. 145–178.

Kleeman, L. (1999). Fast and accurate sonar trackers using double pulse coding, in *1999 IEEE International Conference on Intelligent Robots and Systems*, pp. 1185–1190.

Kleeman, L. and Kuc, R. (1994). An optimal sonar array for target localization and classification, in *IEEE International Conference on Robotics and Automation, 8-13 May 1994*, Vol. 4, pp. 3130–3135.

Kleeman, L. and Kuc, R. (1995). Mobile robot sonar for target localization and classification, *International Journal of Robotic Research* **14**, 4, pp. 295–318.

Kocarev, L., Halle, K., Eckert, K. and Chua, L. O. (1993). Experimental observations of antimonotonicity in chua's circuit, *International Journal of Bifurcations and Chaos* **3**, pp. 1051–1055.

Kocarev, L., Halle, K. S., Eckert, K., Chua, L. . and Parlitz, U. (1992). Experimental demonstration of secure communications via chaotic synchronization, *Int. J. Bifurcation and Chaos* **2**, 3, pp. 709–713.

Lakshmanan, M. and Murali, K. (1995). Experimental chaos from nonautonomous electronic circuits, *Philosophical Transactions: Physical Sciences and Engineering* **353**, 1701, pp. 33–46.

Lee, B.-C., Lee, H.-H. and Wang, B.-H. (1997). Control bifurcation structure of return map control in chua's circuit, *International Journal of Bifurcation and Chaos* **7**, 4, pp. 903–909.

Li, C. and Liao, X. (2004). Lag synchronization of rossler system and chua circuit via a scalar signal, *Physics Letters A* **329**, 4-5, pp. 301–308.

Li, Z. G., Wen, C. Y., Soh, Y. C. and Xie, W. X. (2001). The stabilization and synchronization of chua's oscillators via impulsive control, *IEEE Transactions on circuits and systems I* **48**, 11, pp. 1351–1355.

Lian, K.-Y., Peter, L., Wu, T.-C. and Lin, W.-C. (2002). Chaotic control using fuzzy model-based methods, *International Journal of Bifurcation and Chaos* **12**, 8, pp. 1827–1841.

Liu, Z. (2001). Strange nonchaotic attractors from periodically excited chua's circuit, *International Journal of Bifurcation and Chaos* **11**, 1, pp. 225–230.

Lü, J., Chen, G., Yu, X. and Leung, H. (2004a). Design and analysis of multiscroll chaotic attractors from saturated function series, *IEEE Transactions on Circuits and Systems I* **51**, pp. 2476–2490.

Lü, J., Han, F., Yu, X. and Chen, G. (2004b). Generating 3-d multiscroll chaotic attractors: A hysteresis series switching method, *Automatica* **40**, pp. 1677–1687.

Lukin, K. A. (1993). High frequency oscillation from chua's circuit, *Journal of Circuits and Systems Computation* **3**, pp. 627–643.

Madan, R. (ed.) (1993). *Chua's Circuit: A Paradigm for Chaos* (World Scientific).

Maggio, G. M., Rulkov, N., Sushchik, M., Tsimring, L., Volkovskii, A., Abarbanel, H., Larson, L. and Yao, K. (1999). Chaotic pulse-position modulation for ultrawide-band communication systems, in *Proc. UWB'99, Washington D.C., Sept 28-30*.

Manganaro, G., Arena, P. and Fortuna, L. (1999). *Cellular Neural Networks: Chaos, Complexity and VLSI processing* (Springer-Verlag).

Matsumoto, T., Chua, L. O. and Komuro, M. (1985). The double scroll, *IEEE Transactions on Circuits and Systems* **32**, 8, pp. 798–818.

Matsumoto, T., Chua, L. O. and Tokumasu, K. (1986). Double scroll via a two-transistor circuit, *IEEE Transactions on Circuits and Systems I* **33**, 8, pp. 828–835.

Mayer-Kress, G., Choi, I., Weber, N., Bargar, R. and Hubler, A. (1993). Musical signals from chua's circuit, *IEEE Transactions on Circuits and Systems II* **40**, pp. 688–695.

Munuzuri, A. P., Perez-Munuzuri, V., Gomez-Gesteira, L. O., M.and Chua and Perez-Villar, V. (17–50). Spatiotemporal structures in discretely-coupled arrays of nonlinear circuits: a review, *Int. J. Bifurcation and Chaos* **5**, 1, p. 1995.

Murali, K. and Lakshmanan, M. (1992). Effect of sinusoidal excitation on the chua's circuit, *IEEE Transactions on Circuits and Systems I* **39**, 4, pp. 264–270.

Murali, K. and Lakshmanan, M. (1993). Chaotic dynamics of the driven chua's circuit, *IEEE Transactions on Circuits and Systems I* **40**, 1, pp. 836–840.

Murali, K., Lakshmanan, M. and Chua, L. (1994a). Bifurcation and chaos in the simplest dissipative non-autonomous circuit, *International Journal of Bifurcation and Chaos* **4**, pp. 1511–1524.

Murali, K., Lakshmanan, M. and Chua, L. O. (1994b). The simplest dissipative non-autonomous chaotic circuit, *IEEE Transactions on Circuits and Systems I* **41**, pp. 462–463.

Muscato, G. (2004). The collective behavior of piezoelectric walking microrobots: Experimental results, *Intell. Autom. Soft Comput.* **10**, pp. 267–276.

Nekorkin, V. I. and Chua, L. O. (1993). Spatial disorder and wave fronts in a chain of coupled chua's circuit, *International Journal of Bifurcation and Chaos* **3**, 5, pp. 1281–1291.

Nekorkin, V. I., Kazantsev, V. B. and Chua, L. O. (1996a). Chaotic attractors and waves in a one-dimensional array of modified chua's circuits, *International Journal of Bifurcation and Chaos* **6**, 7, pp. 1295–1317.

Nekorkin, V. I., Kazantsev, V. B., Rulkov, M. F., Velarde, M. G. and Chua, L. . (1995). Homoclinic orbits and solitary waves in a one-dimensional array of chua's circuits, *IEEE Transactions on Circuits and Systems I* **42**, 10, pp. 785–801.

Nekorkin, V. I., Kazantsev, V. B. and Velarde, M. G. (1996b). Travelling waves in a circular array of chua's circuits, *International Journal of Bifurcation and Chaos* **6**, 3, pp. 473–484.

Nossek, J. A. (1994). Design and learning with cellular neural networks, in *Proc. of IEEE Workshop on CNNs and their App.s '94*, pp. 137–146.

Nossek, J. A., Magnussen, P., H.and Nachbar and Schuler, A. J. (1993). Learning algorithms for cellular neural networks, in *Proc. of NOLTA '93*, pp. 11–16.

Ogorzalek, M. J. (1993a). Taming chaos: Part i synchronization, *IEEE Transactions on Circuits and Systems I* **40**, 10, pp. 693–699.

Ogorzalek, M. J. (1993b). Taming chaos: Part ii control, *IEEE Transactions on Circuits and Systems I* **40**, 10, pp. 700–706.

Ogorzalek, M. J. (1995). Controlling chaos in electronic circuits, *Philosophical Transactions: Physical Sciences and Engineering* **353**, 1701, pp. 127–136.

Ott, E., Grebogi, C. and Yorke, J. (1990). Controlling chaos, *Physical Review Letters* **64**, pp. 1196–1199.

Ozoguz, S., Elwakil, A. S. and Salama, K. N. (2002). n-scroll chaos generator using nonlinear transconductor, *Electronics Letters* **38**, pp. 685–686.

Papoulis, A. (1991). *Probability, Random Variables and Stochastic Processes* (McGraw-Hill).

Pecora, L. and Carroll, T. (1998). Master stability functions for synchronized coupled systems, *Physical Review Letters* **80**, pp. 2109–2112.

Pecora, L. M. and Carroll, T. L. (1990). Synchronization in chaotic systems, *Physical Review Letters* **64**, 8, pp. 821–824.

Perez-Munuzuri, A., Perez-Munuzuri, V., Gomez-Gesteria, M., Chua, L. O. and Perez-Villar, V. (1995). Spatiotemporal structures in discretely-coupled arrays of nonlinear circuits: a review, *International Journal of Bifurcation and Chaos* **5**, 1, pp. 17–50.

Perez-Munuzuri, A., Perez-Munuzuri, V. and Perez-Villar, V. (1993a). Spiral waves on a two-dimensional array of nonlinear circuits, *IEEE Transactions on Circuits and Systems I* **40**, pp. 872–877.

Perez-Munuzuri, A., Perez-Villar, V. and Chua, L. O. (1993b). Autowaves for image processing on a two-dimensional cnn array of excitable nonlinear circuits: flat and wrinkled labyrinths, *IEEE Transactions on Circuits and Systems I* **40**, pp. 174–193.

Perez-Munuzuri, V., Gomez-Gesteira, V., Perez-Villar, V. and Chua, L. O. (1993c). Traveling wave propagation in a one-dimensional fluctuating medium, *International Journal of Bifurcation and Chaos* **3**, pp. 211–215.

Perez-Munuzuri, V., Perez-Villar, V. and Chua, L. O. (1992). Propagation failure in linear arrays of chua's circuits, *Int. J. Bifurcation and Chaos* **2**, 2, pp. 403–406.

Perez-Munuzuri, V., Perez-Villar, V. and Chua, L. O. (1993d). Travelling wave front and its failure in a one-dimensional array of chua's circuits, *Journal of Circuits and Systems Computation* **3**, pp. 215–229.

Piccardi, C. and Rinaldi, S. (2002). Control of complex peak-to-peak dynamics, *International Journal of Bifurcation and Chaos* **12**, 12, pp. 2927–2936.

Pikovsky, A., Rosenblum, M. and Kurths, J. (2001). *Synchronization: A universal concept in nonlinear sciences* (Cambridge Nonlinear Science Series 12).

Pinkney, J. Q., Camwell, P. L. and Davies, R. (1995). Chaos shift keying communications using self-synchronizing chua oscillators, *Electronic Letters* **31**, 13, pp. 1021–1022.

Pivka, L. and Spany, V. (1993). Boundary surfaces and basin bifurcations in chua's circuit, *J Cir. Syst. Comput.* **3**, pp. 441–470.

Polaroid (1990). *6500-Series Sonar Ranging Module* (Product Specifications PID 615077, Polaroid Corporation, Cambridge, MA).

Politis, Z. and Probert, P. (1998). Perception of an indoor robot workspace by using ctfm sonar imaging, in *998 IEEE International Conference on Robotics and Automation, Leuven, Belgium, 1998.*

Puebla, H., Alvarez-Ramirez, J. and Cervantes, I. (2003). A simple tracking control for chua's circuit, *IEEE Transactions on circuits and systems I* **50**, 2, pp. 280–284.

Pyragas, K. (1992). Continuous control of chaos by self-controlling feedback, *Physics Letters A* **170**, pp. 421–428.

Pyragas, K. (1995). Control of chaos via extended delay feedback, *Physics Letters A* **206**, pp. 323–330.

Pyragas, K. (2006). Delayed feedback control of chaos, *Philosophical transactions of the royal society A* **364**, pp. 2309–2334.

Pyragas, K. and Tamasevicius, A. (1993). Experimental control of chaos by delayed self-controlling feedback, *Physics Letters A* **180**, pp. 99–102.

Rahma, F., Fortuna, L. and Frasca, M. (2009). New attractors and new behaviours in the photo-controlled Chua's circuit, *Int. J. Bif. Chaos*, to appear.

Rodet, X. (1993a). Models of musical instruments from chua's circuit with time delay, *IEEE Transactions on Circuits and Systems II* **40**, pp. 696–701.

Rodet, X. (1993b). Sound and music from chua's circuit, *Journal of Circuits, Systems, and Computers* **3**, 1, pp. 49–61.

Rodriguez-Vazquez, A. and Delgado-Restituto, M. (1993). Cmos design of chaotic oscillators using state variables: A monolithic chua's circuit, *IEEE Transactions on Circuits and Systems II* **40**, 10, pp. 596–613.

Sanchez, E., Matias, M. A. and Perez-Munuzuri, V. (2000). Chaotic synchronization in small assemblies of driven chua's circuits, *IEEE Transactions on Circuits and Systems I* **47**, 5, pp. 644–654.

Sharkovsky, A. N., Maistrenko, Y., Deregel, P. and Chua, L. O. (1993). Dry turbolence from a time-delayed chua's circuit, *Journal of Circuits and Systems Computation* **3**, pp. 645–668.

Shi, Z. and Ran, L. (2004). Tunnel diode based chua's circuit, in *IEEE 6th CAS Symp. on Emerging Technologies: Mobile and Wireless Comm., Shangai, China, May 31-June2*, pp. 217–220.

Shilnikov, L. P. (1965). A case of the existence of a denumerable set of periodic motions, *Sov. Math. Dokl.* **6**, pp. 163–166.

Strogatz, S. H. (1994). *Nonlinear Dynamics and Chaos* (Perseus Books).

Strukov, D. B., Snider, G. S., Stewart, D. R. and Williams, R. S. (2008). The missing memristor found, *Nature* **453**, pp. 80–83.

Suykens, J. A. K. and Chua, L. O. (1997). n-double scroll hypercubes in 1-d cnns, *International Journal of Bifurcation and Chaos* **7**, 8, pp. 1873–1885.

Suykens, J. A. K., Huang, A. and Chua, L. O. (1997). A family of n-scroll attractors from a generalized chuas circuit, *Archiv fur Elektronik und Ubertragungstechnik* **51**, 3, pp. 131–138.

Suykens, J. K. and Vandewalle, J. (1993). Generation of n-double scrolls (n=1;2;3;4; ...), *IEEE Transactions on Circuits and Systems I* **40**, pp. 861–867.

Tang, K. S., Zhong, G. Q., Chen, G. and Man, K. F. (2001). Generation of n-scroll attractors via sine function, *IEEE Transactions on Circuits and Systems I* **48**, pp. 1369–1372.

Tang, Y. S., Mees, A. I. and Chua, L. O. (1983). Synchronisation and chaos, *IEEE Trans. on Circuit and Systems* **30**, 9, pp. 620–626.

Tonelli, R. and Meloni, F. (2002). Chua's periodic table, *International Journal of Bifurcation and Chaos* **12**, 7, pp. 1451–1464.

Torres, L. and Aguirre, L. (1999). Extended chaos control method applied to chua circuit, *Electronics Letters* **35**, 10, pp. 768–770.

Torres, L. A. B. and Aguirre, L. A. (2000). Inductorless chua's circuit, *Electronics Letters* **36**, 23, pp. 1915–1916.

Tour, J. M. and He, T. (2008). The fourth element, *Nature* **453**, pp. 42–43.

Wang, S.-H., Liu, W.-Q., Ma, B.-J., Xiao, J.-H. and Jiang, D. Y. (2005). Phase synchronization of chua circuit induced by the periodic signals, *Chinese Physics* **14**, 1, pp. 55–60.

Wang, X., Zhong, G., Tang, K.-S., Man, K. and Liu, Z.-F. (2001). Generating chaos in chuas circuit via time-delay feedback, *IEEE Transactions on Circuits and Systems I* **48**, 9, pp. 1151–1156.

Woodward, P. M. (1964). *Probability and Information Theory with Application to Radar* (Oxford Pergamon Press).

Wu, X. Y., Chen, H. and Tan, W. S. V. (1993). The improvement of the hardware design of artificial cnn and a fast algorithm of (cnn) component detector, *J. of Franlin Institute* **330**, 6, pp. 1005–1015.

Xu, S. (1987). Chua's circuit family, *Proceedings of IEEE* **75**, 8, pp. 1022–1032.

Yalcin, M. E., Suykens, J. A. K. and Vandewalle, J. (2000). Experimental confirmation of 3- and 5-scroll attractors from a generalized chua's circuit, **47**, 3, pp. 425–429.

Yalcin, M. E., Suykens, J. A. K., Vandewalle, J. and Ozoguz, S. (2002). Families of scroll grid attractors, *International Journal of Bifurcation and Chaos* **12**, pp. 23–41.

Yang, L. and Liao, Y. (1987). Self-similar bifurcation structures from chuas circuit, *Int. J. Circ. Th. Appl.* **15**, pp. 189–192.

Yang, T. and Chua, L. O. (1996). Secure communication via chaotic parameter modulation, *IEEE Transactions on Circuits and Systems I* **43**, 9, pp. 817–819.

Yang, T. and Chua, L. O. (1997). Impulsive stabilization for control and synchronization of chaotic systems: theory and applications to secure communication, *IEEE Transactions on circuits and systems I* **44**, 10, pp. 976–988.

Yang, T. and Chua, L. O. (2001). Testing for local activity and edge of chaos, *International Journal of Bifurcation and Chaos* **11**, 6, pp. 1495–1591.

Yang, T., Wu, C. W. and Chua, L. O. (1997). Cryptography based on chaotic systems, *IEEE Transactions on circuits and systems I* **44**, 5, pp. 469–472.

Yin, X. and Cao, Y. J. (2003). Sychronisation of chua's oscillator via the state observer technique, *International Journal of Electrical Engineering Education* **40**, 1, pp. 36–44.

Zhang, T. and Feng, G. (2007). Output tracking of piecewise-linear systems via error feedback regulator with application to synchronization of nonlinear chua's circuit, *IEEE Transactions on Circuits and Systems I* **54**, 8, pp. 1852–1863.

Zhong, G. Q. (1994). Implementation of chua's circuit with a cubic nonlinearity, *IEEE Transactions on Circuits and Systems I* **41**, 12, pp. 934–941.

Index of the implementations

Index